VOLUME ONE HUNDRED AND NINETY THREE

METHODS IN CELL BIOLOGY

The Immunological Synapse - Part C

SERIES EDITOR

Lorenzo Galluzzi
Fox Chase Cancer Center,
Philadelphia, PA, United States

VOLUME ONE HUNDRED AND NINETY THREE

METHODS IN CELL BIOLOGY

The Immunological Synapse - Part C

Edited by

CLÉMENT THOMAS
*Luxembourg Institute of Health,
Luxembourg, Luxembourg*

LORENZO GALLUZZI
*Fox Chase Cancer Center,
Philadelphia, PA, United States*

Academic Press is an imprint of Elsevier
125 London Wall, London EC2Y 5AS, United Kingdom
525 B Street, Suite 1650, San Diego, CA 92101, United States
50 Hampshire Street, 5th Floor, Cambridge, MA 02139, United States

First edition 2025

Copyright © 2025 Elsevier Inc. All rights are reserved, including those for text and data mining, AI training, and similar technologies.

Publisher's note: Elsevier takes a neutral position with respect to territorial disputes or jurisdictional claims in its published content, including in maps and institutional affiliations.

No part of this publication may be reproduced or transmitted in any form or by any means, electronic or mechanical, including photocopying, recording, or any information storage and retrieval system, without permission in writing from the publisher. Details on how to seek permission, further information about the Publisher's permissions policies and our arrangements with organizations such as the Copyright Clearance Center and the Copyright Licensing Agency, can be found at our website: www.elsevier.com/permissions.

This book and the individual contributions contained in it are protected under copyright by the Publisher (other than as may be noted herein).

Notices

Knowledge and best practice in this field are constantly changing. As new research and experience broaden our understanding, changes in research methods, professional practices, or medical treatment may become necessary.

Practitioners and researchers must always rely on their own experience and knowledge in evaluating and using any information, methods, compounds, or experiments described herein. In using such information or methods they should be mindful of their own safety and the safety of others, including parties for whom they have a professional responsibility.

To the fullest extent of the law, neither the Publisher nor the authors, contributors, or editors, assume any liability for any injury and/or damage to persons or property as a matter of products liability, negligence or otherwise, or from any use or operation of any methods, products, instructions, or ideas contained in the material herein.

ISBN: 978-0-443-21868-2
ISSN: 0091-679X

For information on all Academic Press publications
visit our website at https://www.elsevier.com/books-and-journals

Publisher: Zoe Kruze
Editorial Project Manager: Devwart Chauhan
Production Project Manager: A. Maria Shalini
Cover Designer: Arumugam Kothandan

Typeset by STRAIVE, India

Contents

Contributors   *xi*

1. Analysis of immune synapses by τau-STED imaging and 3D-quantitative colocalization of lytic granule markers   1

Emilia Scharrig, Maria L. Sanmillan, and Claudio G. Giraudo

1. Introduction	2
2. Materials	4
3. Methods	5
4. Concluding remarks	10
5. Notes	11
Acknowledgments	12
Declaration of interests	12
References	12

2. Imaging the immune synapse: Three-dimensional analysis of the immune synapse   15

Javier Ruiz-Navarro, Sofía Blázquez-Cucharero, Víctor Calvo, and Manuel Izquierdo

1. Introduction	16
2. Materials, cells, immunological synapse formation and image capture	20
3. 3D analysis of the IS	21
4. Discussion, future perspectives and concluding remarks	31
5. Notes	33
Conflict of interest	34
Author contributions	34
Acknowledgments	35
References	35

3. Image processing approaches for microtubule remodeling quantification at the immunological synapse   39

Daniel Krentzel, Maria Isabella Gariboldi, Marie Juzans, Marta Mastrogiovanni, Florian Mueller, Céline Cuche, Vincenzo Di Bartolo, and Andrés Alcover

1. Introduction	41

2. Materials	44
3. Methods	49
4. Results and discussion	60
5. Concluding remarks	64
6. Notes	64
Acknowledgments	65
Declaration of interests	66
References	66

4. A comprehensive guide to study the immunological synapse using imaging flow cytometry — 69

Andrea Michela Biolato, Liza Filali, Max Krecke, Clément Thomas, and Céline Hoffmann

1. Introduction	70
2. Materials	72
3. Methods	74
4. Concluding remarks	92
5. Notes	92
Acknowledgments	96
Bibliography	96

5. Quantifying force-mediated antigen extraction in the B cell immune synapse using DNA-based tension sensors — 99

Hannah C.W. McArthur, Anna T. Bajur, and Katelyn M. Spillane

1. Introduction	100
2. Materials	102
3. Methods	108
4. Notes	121
5. Conclusions	124
Acknowledgments	124
References	124

6. Gauging antigen recognition by human primary T-cells featuring orthotopically exchanged TCRs of choice — 127

Vanessa Mühlgrabner, Angelika Plach, Johannes Holler, Judith Leitner, Peter Steinberger, Loïc Dupré, Janett Göhring, and Johannes B. Huppa

1. Introduction	128
2. Material	131
3. Methods	134

4. Notes	150
5. Conclusion	151
Acknowledgments	152
References	152
Further reading	154

7. Measuring interaction kinetics between T cells and their target tumor cells with optical tweezers — **155**

Edison Gerena, Sophie Goyard, Nicolas Inacio, Jerko Ljubetic, Amandine Schneider, Sinan Haliyo, and Thierry Rose

1. Introduction	156
2. Material	161
3. Methods	163
4. Results	168
5. Concluding remarks	169
6. Perspectives	171
7. Notes	171
Acknowledgments	172
Declaration of interest	172
References	172

8. Measuring interaction force between T lymphocytes and their target cells using live microscopy and laminar shear flow chambers — **175**

Sophie Goyard, Amandine Schneider, Jerko Ljubetic, Nicolas Inacio, Marie Juzans, Céline Cuche, Pascal Bochet, Vincenzo Di Bartolo, Andrés Alcover, and Thierry Rose

1. Introduction	176
2. Material	179
3. Methods	181
4. Results	192
5. Concluding remarks	194
6. Perspectives	194
Notes	195
Ethics statement	197
Acknowledgments	197
Declaration of interest	198
References	198

viii

Contents

9. Isolation and characterization of primary NK cells and the enrichment of the KIR2DL1$^+$ population 201

Batel Sabag, Abhishek Puthenveetil, and Mira Barda-Saad

1. Introduction	202
2. Materials	203
3. Methods	205
4. Concluding remarks	210
Acknowledgments	211
References	211

10. Flow cytometry conjugate formation assay between natural killer cells and their target cells 213

Gilles Iserentant, Carole Seguin-Devaux, and Jacques Zimmer

1. Introduction	214
2. Materials	216
3. Methods	217
4. Results	219
5. Discussion	225
References	227

11. Quantitative analysis of the B cell immune synapse using imaging techniques 229

Oreste Corrales Vázquez, Teemly Contreras, Martina Alamo Rollandi,

Felipe Del Valle Batalla, and Maria-Isabel Yuseff

1. Introduction	230
2. Materials and methods	230
3. Analysis of B cell activation with immobilized antigens	234
4. Analysis of the distribution of organelles in B cells activated with antigen-coated beads	236
5. Characterization of the immune synapse organization of B cells activated on antigen-coated surfaces	244
6. Discussion and relevance	250
7. Notes	250
References	252

12. γδ T cell expansion and their use in *in vitro* cytotoxicity assays   253

María Alejandra Parigiani

1. Introduction	254
2. Graphical abstract	256
3. Materials	256
4. Methods	258
5. Concluding remarks	262
6. Notes	262
Acknowledgments	263
Declaration of interests	263
References	263

Contributors

Martina Alamo Rollandi
Laboratory of Immune Cell Biology, Department of Cellular and Molecular Biology, Pontificia Universidad Católica de Chile, Santiago, Chile

Andrés Alcover
Institut Pasteur, Université Paris Cité, INSERM-U1224, Unité Biologie Cellulaire des Lymphocytes, Ligue Nationale Contre le Cancer, Équipe Labellisée Ligue-2018, Paris, France

Anna T. Bajur
Department of Physics; Randall Centre for Cell & Molecular Biophysics, King's College London, London, United Kingdom

Mira Barda-Saad
The Mina and Everard Goodman Faculty of Life Sciences, Bar-Ilan University, Ramat-Gan, Israel

Andrea Michela Biolato
Cytoskeleton and Cancer Progression, Department of Cancer Research, Luxembourg Institute of Health, Luxembourg; Faculty of Science, Technology and Medicine, University of Luxembourg, Esch-sur-Alzette, Luxembourg

Sofía Blázquez-Cucharero
Instituto de Investigaciones Biomédicas Alberto Sols CSIC-UAM, Madrid, Spain

Pascal Bochet
Institut Pasteur, Université de Paris Cité, Image Analysis Unit, CNRS UMR3691, Paris, France

Víctor Calvo
Departamento de Bioquímica, Instituto de Investigaciones Biomédicas Alberto Sols CSIC-UAM, Facultad de Medicina, Universidad Autónoma de Madrid, Madrid, Spain

Teemly Contreras
Laboratory of Immune Cell Biology, Department of Cellular and Molecular Biology, Pontificia Universidad Católica de Chile, Santiago, Chile

Oreste Corrales Vázquez
Laboratory of Immune Cell Biology, Department of Cellular and Molecular Biology, Pontificia Universidad Católica de Chile, Santiago, Chile

Céline Cuche
Institut Pasteur, Université Paris Cité, INSERM-U1224, Unité Biologie Cellulaire des Lymphocytes, Ligue Nationale Contre le Cancer, Équipe Labellisée Ligue-2018, Paris, France

Felipe Del Valle Batalla
Laboratory of Immune Cell Biology, Department of Cellular and Molecular Biology, Pontificia Universidad Católica de Chile, Santiago, Chile

Vincenzo Di Bartolo
Institut Pasteur, Université Paris Cité, INSERM-U1224, Unité Biologie Cellulaire des Lymphocytes, Ligue Nationale Contre le Cancer, Équipe Labellisée Ligue-2018, Paris, France

Loïc Dupré
Medical University of Vienna, University Clinics for Dermatology, Vienna, Austria

Liza Filali
Cytoskeleton and Cancer Progression, Department of Cancer Research, Luxembourg Institute of Health, Luxembourg, Luxembourg

Maria Isabella Gariboldi
Institut Pasteur, Université Paris Cité, CNRS-UMR3691, Unité Imagerie et Modélisation, Paris, France

Edison Gerena
Institut des Systèmes Intelligents et de Robotique (ISIR), Sorbonne Université, CNRS, Paris, France

Claudio G. Giraudo
Department of Microbiology and Immunology- Sydney Kimmel Medical College- Thomas Jefferson University, Philadelphia, PA, United States

Janett Göhring
Medical University of Vienna, Center for Pathophysiology, Infectiology and Immunology, Institute for Hygiene and Applied Immunology, Vienna, Austria

Sophie Goyard
Institut Pasteur, Université Paris Cité, Diagnostic Test Innovation & Development Core Facility, Paris, France

Sinan Haliyo
Institut des Systèmes Intelligents et de Robotique (ISIR), Sorbonne Université, CNRS, Paris, France

Céline Hoffmann
Cytoskeleton and Cancer Progression, Department of Cancer Research, Luxembourg Institute of Health, Luxembourg; National Cytometry Platform, Luxembourg Institute of Health, Esch-sur-Alzette, Luxembourg

Johannes Holler
Medical University of Vienna, Center for Pathophysiology, Infectiology and Immunology, Institute for Hygiene and Applied Immunology, Vienna, Austria

Johannes B. Huppa
Medical University of Vienna, Center for Pathophysiology, Infectiology and Immunology, Institute for Hygiene and Applied Immunology, Vienna, Austria

Nicolas Inacio
Institut des Systèmes Intelligents et de Robotique (ISIR), Sorbonne Université, CNRS; Institut Pasteur, Université Paris Cité, Diagnostic Test Innovation & Development Core Facility, Paris, France

Gilles Iserentant
Department of Infection and Immunity, Luxembourg Institute of Health, Esch-sur-Alzette, Luxembourg

Manuel Izquierdo
Instituto de Investigaciones Biomédicas Alberto Sols CSIC-UAM, Madrid, Spain

Marie Juzans
Institut Pasteur, Université Paris Cité, INSERM-U1224, Unité Biologie Cellulaire des Lymphocytes, Ligue Nationale Contre le Cancer, Équipe Labellisée Ligue-2018; Sorbonne Université Collège Doctoral, Paris, France

Max Krecke
Cytoskeleton and Cancer Progression, Department of Cancer Research, Luxembourg Institute of Health, Luxembourg; Faculty of Science, Technology and Medicine, University of Luxembourg, Esch-sur-Alzette, Luxembourg

Daniel Krentzel
Institut Pasteur, Université Paris Cité, CNRS-UMR3691, Unité Imagerie et Modélisation, Paris, France

Judith Leitner
Medical University of Vienna, Center for Pathophysiology, Infectiology and Immunology, Institute for Immunology, Vienna, Austria

Jerko Ljubetic
Institut Pasteur, Université Paris Cité, INSERM-U1224, Unité Biologie Cellulaire des Lymphocytes, Ligue Nationale Contre le Cancer, Équipe Labellisée Ligue-2018; Institut Pasteur, Université Paris Cité, Diagnostic Test Innovation & Development Core Facility, Paris, France

Marta Mastrogiovanni
Institut Pasteur, Université Paris Cité, INSERM-U1224, Unité Biologie Cellulaire des Lymphocytes, Ligue Nationale Contre le Cancer, Équipe Labellisée Ligue-2018; Sorbonne Université Collège Doctoral, Paris, France

Hannah C.W. McArthur
Department of Physics, King's College London, London, United Kingdom

Florian Mueller
Institut Pasteur, Université Paris Cité, CNRS-UMR3691, Unité Imagerie et Modélisation, Paris, France

Vanessa Mühlgrabner
Medical University of Vienna, Center for Pathophysiology, Infectiology and Immunology, Institute for Hygiene and Applied Immunology, Vienna, Austria

María Alejandra Parigiani
Faculty of Biology; Signalling Research Centres BIOSS and CIBSS, University of Freiburg; Center of Chronic Immunodeficiency (CCI) and Institute for Immunodeficiency, University Clinics and Medical Faculty, Freiburg, Germany

Angelika Plach
Medical University of Vienna, Center for Pathophysiology, Infectiology and Immunology, Institute for Hygiene and Applied Immunology, Vienna, Austria

Abhishek Puthenveetil
The Mina and Everard Goodman Faculty of Life Sciences, Bar-Ilan University, Ramat-Gan, Israel

Thierry Rose
Institut Pasteur, Université Paris Cité, Diagnostic Test Innovation & Development Core Facility; Institut Pasteur, Université Paris Cité, INSERM-U1224, Unité Biologie Cellulaire des Lymphocytes, Ligue Nationale Contre le Cancer, Équipe Labellisée Ligue-2018, Paris, France

Javier Ruiz-Navarro
Instituto de Investigaciones Biomédicas Alberto Sols CSIC-UAM, Madrid, Spain

Batel Sabag
The Mina and Everard Goodman Faculty of Life Sciences, Bar-Ilan University, Ramat-Gan, Israel

Maria L. Sanmillan
Department of Microbiology and Immunology- Sydney Kimmel Medical College- Thomas Jefferson University, Philadelphia, PA, United States

Emilia Scharrig
Department of Microbiology and Immunology- Sydney Kimmel Medical College- Thomas Jefferson University, Philadelphia, PA, United States

Amandine Schneider
Institut Pasteur, Université Paris Cité, INSERM-U1224, Unité Biologie Cellulaire des Lymphocytes, Ligue Nationale Contre le Cancer, Équipe Labellisée Ligue-2018; Institut Pasteur, Université Paris Cité, Diagnostic Test Innovation & Development Core Facility, Paris, France

Carole Seguin-Devaux
Department of Infection and Immunity, Luxembourg Institute of Health, Esch-sur-Alzette, Luxembourg

Katelyn M. Spillane
Department of Physics; Randall Centre for Cell & Molecular Biophysics, King's College London; Department of Life Sciences, Imperial College London, London, United Kingdom

Peter Steinberger
Medical University of Vienna, Center for Pathophysiology, Infectiology and Immunology, Institute for Immunology, Vienna, Austria

Clément Thomas
Cytoskeleton and Cancer Progression, Department of Cancer Research, Luxembourg Institute of Health, Luxembourg, Luxembourg

Maria-Isabel Yuseff
Laboratory of Immune Cell Biology, Department of Cellular and Molecular Biology, Pontificia Universidad Católica de Chile, Santiago, Chile

Jacques Zimmer
Department of Infection and Immunity, Luxembourg Institute of Health, Esch-sur-Alzette, Luxembourg

CHAPTER ONE

Analysis of immune synapses by τau-STED imaging and 3D-quantitative colocalization of lytic granule markers

Emilia Scharrig, Maria L. Sanmillan, and Claudio G. Giraudo*

Department of Microbiology and Immunology- Sydney Kimmel Medical College- Thomas Jefferson University, Philadelphia, PA, United States
*Corresponding author: e-mail address: claudio.giraudo@jefferson.edu

Contents

1. Introduction	2
2. Materials	4
2.1 Common disposables	4
2.2 Equipment	4
2.3 Reagents	4
2.4 Cytotoxic cells and target cells culture	5
2.5 Data analysis	5
3. Methods	5
3.1 Cell culture	5
3.2 Cell conjugation	6
3.3 Immunofluorescence	6
3.4 τau-STED microscopy imaging	7
3.5 Image analysis	8
4. Concluding remarks	10
5. Notes	11
Acknowledgments	12
Declaration of interests	12
References	12

Abstract

Over the last decades, intensive research studies have been focused on describing how the immunological synapse is formed, the intracellular mechanisms that control lytic granules formation, and even further, the steps toward granule polarization before the killing event is achieved. These convoluted processes pose significant experimental challenges since the components' sizes are smaller than the diffraction limit of the

---

Methods in Cell Biology, Volume 193
ISSN 0091-679X
https://doi.org/10.1016/bs.mcb.2023.01.018

Copyright © 2025 Elsevier Inc.
All rights are reserved, including those
for text and data mining, AI training,
and similar technologies.

conventional fluorescent microscopy techniques and their highly dynamic nature. Here, we describe a procedure to perform a quantitative analysis of the protein markers of these lytic granules by using τau-STED imaging and 3D-quantitative colocalization of lytic granule markers. The innovative technology offered by τau-STED microscopy and unbiased imaging analysis is a great tool that could be applied to further our understanding of lytic granule composition and localization and study other dynamic processes at the immunological synapses.

Abbreviations

CTLs	cytotoxic T-lymphocytes
IS	immunological synapse
LG	lytic granules
NK	natural killer
ON	overnight
PBMC	peripheral blood mononuclear cells
PFA	paraformaldehyde
PLL	poly-L-lysine
RT	room temperature
TCR	T-cell receptor

1. Introduction

The immunological synapse (IS) is an architecturally well-defined structure formed at the interface of cytotoxic T-lymphocytes (CTLs) or natural killer (NK) cells upon recognition of foreign antigens presented on the cell surface of a virally infected or cancer cell (a.k.a. target cells) (Dustin & Long, 2010; McCann et al., 2003; Stinchcombe, Bossi, Booth, & Griffiths, 2001). CTLs and NK cells utilize the T-Cell Receptor (TCR) or other activating receptors, respectively, to recognize foreign antigens on target cells and trigger signaling pathways that ultimately lead to the killing of target cells. Target cell death is mediated by the polarized exocytosis of cytolytic proteins such as the pore-forming protein perforin and apoptosis-inducing granzymes contained in specialized lysosome-related organelles known as lytic granules (LGs). Persistent signaling through the TCR or activator receptors is critical to sustaining the formation of a stable IS. At the same time, LGs need to undergo a maturation process by which the LGs merge with specific *endo*-lysosomal compartments to acquire factors necessary for their transport and polarization toward the IS, fusion, and release of their cytolytic content, and induction of apoptosis of the target cell (de Saint, Menasche, & Fischer, n.d.). This highly dynamic and convoluted process of CTLs and NK cells is of the utmost importance for adaptive immunity against viral infections and tumors (Orange & Ballas, n.d.; Cerwenka & Lanier, 2001).

It has been proposed that resting CTLs mainly contain immature LGs bearing perforin, which upon activation, need to undergo a series of maturation steps that involve the fusion with Rab7-containing late endosomes first, followed by a merging with Rab11 positive recycling endosomes (de Saint et al., 2010; Menager et al., 2007). The small size of these organelles and the speed of these intracellular fusion events have imposed a significant challenge for their visualization, identification, and tracking using conventional fluorescent microscopy techniques. In this article, we describe how STimulated Emission Depletion (STED) (Hell & Wichmann, 1994) microscopy can be used to overcome these obstacles.

STED (Hell & Wichmann, 1994) microscopy is a fluorescence microscopy super-resolution technique that can circumvent the optical diffraction limit and achieve super-resolution imaging below the 100 nm range (Meyer et al., 2008). It creates super-resolution images by the selective deactivation of fluorophores, minimizing the area of illumination at the focal point and thus enhancing the achievable resolution for a given system. This technique relies on the overlap of a doughnut-shaped STED beam leading to an increase in resolution while decreasing the background noise. In this way, when STED is applied, the overlapping STED doughnut results in depletion of the fluorophores in the outer region of the excitation spot; this methodical depletion results in the confinement of the area in which the fluorophores are still allowed to fluoresce. The Leica TCS SP8 STED 3× confocal/super-resolution microscope provides a combination of super-resolution in the lateral and in the Z dimension, a significant step forward in imaging resolution capable of achieving 40–60 nm lateral and 130 nm axial resolution. Furthermore, τau-STED microscopy combines the optical signals from STED with the physical information from the fluorescence lifetime acquired at typical confocal speeds (Digman, Dalal, Horwitz, & Gratton, 2008), using phasor analysis in a novel way and works in an automated manner, enabling increased STED resolution up to 30 nm range and elimination of uncorrelated background noise even at low excitation and STED light dose, preserving the delicate structures in the image even at a very low photon budget (Wang et al., 2021), making this the ideal tool for imaging proteins colocalization in small structures as lytic granules.

Here, we provide a detailed experimental protocol for measuring the 3D-colocalization of proteins inside the lytic granule during the immunological synapse in human NK or $CD8^+$ T cells and describe the multi-step approach from cell culture to managing the software. Thus, generating a versatile multi-step approach that can be easy to follow.

2. Materials

2.1 Common disposables

(1) Polystyrene flasks (75 and 25 cm^2) and 24-well plates for cell culture
(2) Automatic pipette and serological pipettes (5 and 10 mL)
(3) Micropipettes (0.2–2 μL, 1–20 μL, 20–200 μL, 200–1000 μL) and tips
(4) Disposable cell counting slides or Neubauer hemocytometer
(5) 1.5 mL microcentrifuge tubes
(6) Glass Coverslips # 1.5, circular 12 mm, thickness 0.13–0.17 mm
(7) Curved very fine precision tip forceps
(8) Parafilm

2.2 Equipment

(1) Incubator (37 °C, 5% CO_2)
(2) Tissue culture centrifuge
(3) Sterile cell culture laminar flow hood safety level II
(4) Leica SP8 3× τau-STED microscope

2.3 Reagents

(1) Cell culture medium RPMI 1640 with GlutaMAX and HEPES (ThermoFisher Scientific, cat# 72400047)
(2) Cell culture medium DMEM (ThermoFisher Scientific, cat# 11966025)
(3) Cell culture medium OpTmizer™ CTS™ T-Cell expansion basal medium (ThermoFisher Scientific, cat# A10221-01)
(4) Fetal Bovine Serum (Seradigm, cat# 97068-091)
(5) Human IL-2 Proleukin (aldesleukin, Clinigen- PA)
(6) CD3/CD28 DynaBeads (Human T-Expander CD3/CD28, Thermofisher Scientific, cat# 11348D)
(7) Phosphate buffer saline (PBS), pH 7.4 (Thermo Scientific, cat# J61196-AP)

(8) Paraformaldehyde solution 4% (Electron Microscopy Sciences, cat# 15700)
(9) Bovine Serum Albumin (EMD- cat# EM-2930)
(10) Triton x-100 (USB cat# 22686)
(11) Mounting media ProLong™ Glass Antifade Mounting (Invitrogen—cat# 36980)
(12) Poly-L-lysine hydrobromide solution 1 mg/mL (Sigma-Aldrich cat# P-2636)
(13) Trypan blue 0.4% solution (Thermofisher Scientific, cat# T10282)
(14) Primary Antibodies: Monoclonal mouse anti-Munc13-4 clone C2 (Santa Cruz, cat #sc-271300), monoclonal rabbit anti-Rab7 clone D95F2 (Cell Signaling, cat# 9367), monoclonal rabbit anti-Rab11a clone D4F5 (Cell Signaling, cat# 5589), mouse monoclonal anti-human Perforin 1, clone DG9 conjugated with Alexa Fluor 647 (BD Biosciences, cat# 556577), monoclonal mouse anti-human CD3 (clone OKT3, eBioscience cat# 16-0037-85)
(15) Secondary antibodies: Alexa Fluor 488 conjugated goat anti-mouse-IgG (Invitrogen, cat# 11029) and Cy3 conjugated donkey anti-rabbit-IgG (Jackson Immune Research, cat# 711-165-152)

2.4 Cytotoxic cells and target cells culture

(1) Primary human CD8+ lymphocytes from healthy donors
(2) Human natural killer cell lymphoblastic leukemia/lymphoma YTS (Gift from Dr. Strominger, Harvard Medical School)
(3) Human HLA-negative B cell lymphoblastoid cell line 721.221 (ATCC CRL-1855)
(4) Murine P815 mastocytoma cell line (ATTC TIB-64)

2.5 Data analysis

(1) ImageJ bundled with Java 1.8.0_172
(2) Scientific Volume Imaging Huygens professional 21.10 software

3. Methods

3.1 Cell culture

Cells are routinely maintained at 37 °C and 5% CO_2 in standard culture conditions. YTS and 721.221 cultured in Roswell Park Memorial Institute

(RPMI) 1640 Medium with GlutaMAX and HEPES, supplemented with 10% fetal bovine serum (FBS) and penicillin and streptomycin (P&S). P815 cells (target cell line for CTLs) are maintained in DMEM supplemented with 10% fetal bovine serum and P&S. CTLs are activated and expanded by culturing them with 1:2 ratio (cell:bead) Dynabeads (Human T-Expander CD3/CD28, Thermofisher) in the presence of 200 mU human IL-2 after the purification of PBMC from a healthy donor, in addition to the standard culture conditions these cells are maintained with a gentle movement to promote growth.

3.2 Cell conjugation
3.2.1 Preparation of coverslips with PLL
Sterile round coverslips were placed in 24-well plates, one per well, and covered with a solution of poly-L-Lysine 1.0 mg/mL in PBS $^{-/-}$ 90 min at RT (Note A). Coverslips are then rinsed three times with PBS and left to air dry in the hood.

5×10^5 effector CD8$^+$ T cells and 5×10^5 target P815 cells were washed three times in PBS to remove the culture medium and mixed in 250 μL of RPMI medium without antibiotics. Before combining effector cells with target cells, target cells were incubated with human anti-CD3 antibody (OKT3) at 5 μg/mL for 15 min. CD8$^+$ T cells were mixed with P815 cells and centrifuged together for 5 min at 400 rpm. Pelleted cells were incubated at 37 °C for 15–20 min to allow conjugation and immune synapse formation. Then conjugated cells are resuspended in media slowly to prevent dissociation of effector and target cells. 70 μL of the cell suspension is added onto each dry PLL-coated coverslip and incubated for 10–15 min at 37 °C to allow cell adhesion to glass. Similar conjugation procedure is used for NK cells and 721.221 target cells, except that the preincubation with OKT3 antibody is not required. Typically, 15–30 min is the optimal time to allow proper immune synapse formation and LG convergence and polarization. 300 μL of freshly prepared PFA 4% was added to each well and incubated at RT 20 min. PFA was then replaced with PBS (Note B).

3.3 Immunofluorescence
Coverslips with attached conjugated cells were washed three times with PBS and then incubated with 0.1% Triton x-100 in PBS for 10 min at RT for proper cell permeabilization. Permeabilization reagent was removed by rinsing three times with PBS and incubated with 3% BSA dissolved in

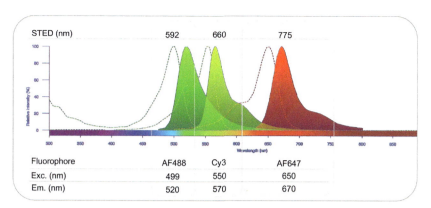

Fig. 1 Fluorescent label combinations and excitation and emission channels used for triple color τau-STED and confocal microscopy.

PBS for 30 min (Note C). Primary antibodies were diluted to the desired concentration in 3% BSA and incubated O.N. at 4 °C in a wet chamber, typically 1:200 dilution. Coverslips were inverted onto a drop of antibody on a strip of parafilm placed in a wet chamber and closed tightly. Cells were washed three times for 5 min each with PBS and incubated with secondary antibodies for 1 h at RT (Fig. 1) (Note D). Cells were washed four times for 5 min each with PBS and rinsed five times in distilled water to clean the excess salts and mounted with Prolong Glass antifade, gently squeezing excess liquid. Mounted coverslips were kept ON in a dark place to cure and reach the correct refractive index (1.52). After that, slides were stored at 4 °C to ensure proper conservation for the secondary fluorophores to last until the observation under the microscope.

3.4 τau-STED microscopy imaging

τau-STED images were acquired on a Leica SP8 tau-STED microscope using an HC PL APO CS2 100×/1.40 oil immersion objective lens (Leica Microsystems, Mannheim, Germany). Excitation was provided by a white light laser (WLL) set at the desired wavelength for each fluorophore. Alexa 488 was excited at 499 nm, Cy3 was excited at 550 nm, and Alexa 647 was excited at 650 nm (Fig. 1). Emission depletion was accomplished with the 592 nm STED laser at 40% maximum power for Alexa 488, 660 STED laser at 25% maximum power for Cy3, and 775 nm STED laser at 30% maximum power for Alexa 647. Fluorescence emission was acquired using the FLIM mode by sequential frame using an independent gated-hybrid detector for each channel (Leica Microsystems). 1024 × 1024-pixel images were acquired with three frame repetitions with a 4.5× of zoom, typically providing a pixel size of 14–29 nm. Confocal images for comparison with the τau-STED images were acquired almost in identical conditions as described above with a Leica SP8 τau-STED microscope in the confocal mode using an HC PL APO CS2 100×/1.40 oil immersion objective lens (Leica Microsystems, Mannheim, Germany). WWL was also used as excitation source as described in Fig. 1. 1024 × 1024-pixel images were acquired with three frame repetitions at 4.5× of zoom, achieving the same pixel size of 14–29 nm as with τau-STED.

3.5 Image analysis

For image analysis, we combined tools from the Huygens software Professional v21.10 and Fiji software to perform a 3-D object-based colocalization analysis (ColocV) which determine the number of voxel (3-D pixel) intersections between two objects. This method allows to quantify colocalization of two markers within the same organelle (vesicle) and on surface contact areas between two different structures. Although the analysis described below can also be performed in other image analysis software, such Imaris (Bitplane), or entirely done in the open-source Fiji/ImageJ, here we described the procedures using Huygens/Fiji which in our hands was more straightforward and provided better results. To do this analysis we performed the following procedure: first the image stack is opened in Huygens, and the image parameters are corrected if necessary. The image is then saved in two different formats, .tif without compression to open it in Fiji and in .h5 to re-open it in Huygens software (Fig. 2B). Once all the images are corrected and deconvolved, to measure the granules content and markers colocalization, the image was first opened in Fiji, following the commands: >image>stacks>z-project. The first and last slices can be chosen at this point, avoiding the extra ones. For clarity purposes, it is better to perform the Max Intensity projection. To assign the desired color to each channel we followed the commands: >image>color>channel tools and selecting the color for each channel of interest. If necessary, split the composite image to have individual-colored images. Use these images as a model to follow for the proper assignment of granule characteristics at Huygens. Next to the image in Fiji that will serve as a model, open in Huygens software the same photo. Follow the commands >Analyze Image>Object analyzer. The segmentation (threshold and seed) is adjusted in each channel following the size of the LG in the .tiff image opened in

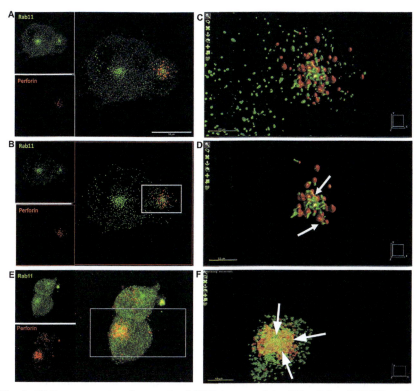

Fig. 2 (A) Representative deconvolved STED image of the NK cell line YTS-WT conjugated with a 721.221 cell. Red Perforin, green Rab11. (B) Same image opened in Huygens software under the Object analyzer. (C) Zoom in of the Region of Interest (ROI) shown in panel (B) as a white rectangle without or with (D) filtering the no overlapping objects; an overlapping volume of Perforin (red vesicles) and Rab11 (green vesicles) are shown in yellow and pointed with white arrows. (E) Representative deconvolved confocal image of the NK cell line YTS-WT conjugated with a 721.221 cell. Red Perforin, green Rab11. (F) Zoom in of the region of interest (ROI) shown in panel (E) as white rectangle with filtering the no overlapping objects.

Fiji. As the images are taken under the same parameters, they usually have very similar values of threshold and seed for all of them. We filter vesicles with more than 100 voxels to perform the analysis by following the commands: filter>remove small objects. Only two channels can be used at the same time. Subsequently, filter non-overlapping objects and select the Region of Interest (ROI), and the command *"keep objects under selected area, discarding others"* is applied. Finally, perform the ColocV analysis to the selection (experiment preset: colocalization—intersection), data from the table analysis (Fig. 3B) is exported and the results plotted using

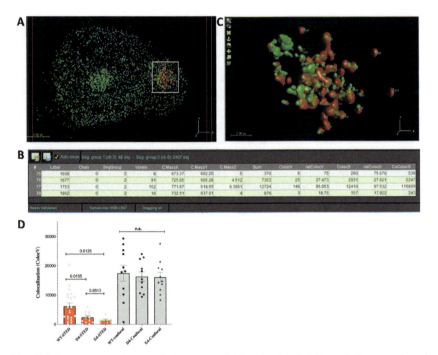

Fig. 3 (A) Representative STED Image YTS Munc13-4 KD (clone D4) opened and deconvolved on Huygens Software. (B) Parameters evaluated of the selected ROI. (C) Zoom in of the ROI shown in panel (A) (white square) with the number object. (D) Graphic showing how higher resolution STED images (red bars), but not confocal images (gray bars) show a decrease in 3D-colocalization of between Rab11 (green) and Perforin (red) in the three different YTS cell lines: wild type (WT), Munc13-4 KD clone D4 (D4) and Munc13-4KD clone E4 (E4) using values of ColocV: voxels intersecting with another object.

Prism software (GraphPad) (Fig. 3C). The differences in the quantification of colocalization of the LGs proteins between a τau–STED image and a normal confocal image are statistically significant (Fig. 3D), supporting the use of this methodology and procedure for the study patterns of protein compartamentalization within individual vesicles during the immunological synapse formation.

4. Concluding remarks

Biological processes that occur during the immunological synapse, such as intracellular granule trafficking, have imposed a challenge for detailed microscopy studies. The nanoscopy techniques and, in particular, the combination of τau–STED with the 3D-colocalization analysis described here allows us to apply quantitative analysis to comprehend the immunological synapse functions better. It can theoretically achieve unlimited resolution, but experimental constraints on biological samples considerably reduce the spatial resolution to about 30 nm in x-y. Moreover, a series of factors related to cell labeling (Lau, Lee, Sahl, Stearns, & Moerner, 2012; Stanly et al., 2016) and image acquisition (Galiani et al., 2012; Hebisch, Wagner, Westphal, Sieber, & Lehnart, 2017; Leutenegger, Eggeling, & Hell, 2010) must be carefully assessed and adjusted depending on the biological process under investigation. Examples of acquisition parameters that must be carefully adjusted in STED microscopy are the STED beam intensity, the excitation beam integration, and the pixel dwell-time. There is a trade-off between several conditions to avoid the onset of unwanted sample degradation effects such as fluorophore photobleaching. This trade-off is often specific to the biological sample considered and cannot be easily determined using calibration samples (i.e., fluorescent spheres). Thus, quantifying the imaging performances directly on the acquired images in an unbiased way is of utmost importance (Mueller et al., 2011; Nieuwenhuizen et al., 2013).

5. Notes

(A) The treatment of the coverslips is performed by submerging its surfaces in the 0.1 mg/mL poly-L-lysine solution at least for 90 min at RT or by incubating at 4 °C ON. The PLL solution could also be performed with a cell culture medium without antibiotics instead of PBS.

(B) Fixed cells could be stored at 4 °C for several days before the immunofluorescence is performed. It is essential to close the plate containing the coverslips with the cells tightly to conserve the humidity and not get dry. It is also recommended to perform only one $PBS^{+/+}$ wash; in this way, the traces of PFA remaining in the solution will help maintain the cell coverslips aseptically.

(C) Cells could be stored in blocking buffer ON at 4 °C if desired or for extended periods. It is recommendable to add sodium azide 0.02% (p/v) to ensure the excellent condition of the blocking buffer.

(D) Up to three fluorescent label combinations are allowed for STED microscopy. A growing list of secondary antibodies can be used in STED microscopy field (Leica Microsystems, 2021). Combinations shown in Fig. 1 were tested for minimal cross-talk in the detection channels for three fluorescent labels.

Acknowledgments

This work was supported by NIH-NIAID-R01AI123538 and R01GM123020 grants to CGG. We would like to thank to the Bioimaging Shared Resource of the Sidney Kimmel Cancer Center (NCI 5 P30 CA-56036).

Declaration of interests

There is no conflict of interest to declared.

References

Cerwenka, A., & Lanier, L. L. (2001). Natural killer cells, viruses and cancer. *Nature Reviews. Immunology, 1*(1), 41–49. Epub 2002/03/22. https://doi.org/10.1038/35095564. PMID: 11905813.

de Saint, B. G., Menasche, G., & Fischer, A. (2010). Molecular mechanisms of biogenesis and exocytosis of cytotoxic granules. *Nature Reviews. Immunology, 10*(8), 568–579. Epub 2010/07/17. doi: nri2803 [pii]10.1038/nri2803. PubMed PMID: 20634814.

Digman, M. A., Dalal, R., Horwitz, A. F., & Gratton, E. (2008). Mapping the number of molecules and brightness in the laser scanning microscope. *Biophysical Journal, 94*(6), 2320–2332. Epub 2007/12/22. doi: 10.1529/biophysj.107.114645. PubMed PMID: 18096627; PubMed Central PMCID: PMCPMC2257897.

Dustin, M. L., & Long, E. O. (2010). Cytotoxic immunological synapses. *Immunological Reviews, 235*(1), 24–34. Epub 2010/06/12. doi: IMR904 [pii] 10.1111/j.0105-2896.2010.00904.x. PubMed PMID: 20536553.

Galiani, S., Harke, B., Vicidomini, G., Lignani, G., Benfenati, F., Diaspro, A., et al. (2012). Strategies to maximize the performance of a STED microscope. *Optics Express, 20*(7), 7362–7374. Epub 2012/03/29. doi: 10.1364/OE.20.007362. PubMed PMID: 22453416.

Hebisch, E., Wagner, E., Westphal, V., Sieber, J. J., & Lehnart, S. E. (2017). A protocol for registration and correction of multicolour STED superresolution images. *Journal of Microscopy, 267*(2), 160–175. Epub 2017/04/04. doi: 10.1111/jmi.12556. PubMed PMID: 28370211.

Hell, S. W., & Wichmann, J. (1994). Breaking the diffraction resolution limit by stimulated emission: Stimulated-emission-depletion fluorescence microscopy. *Optics Letters, 19*(11), 780–782. Epub 1994/06/01. https://doi.org/10.1364/ol.19.000780. PMID: 19844443.

Lau, L., Lee, Y. L., Sahl, S. J., Stearns, T., & Moerner, W. E. (2012). STED microscopy with optimized labeling density reveals 9-fold arrangement of a centriole protein. *Biophysical Journal, 102*(12), 2926–2935. Epub 2012/06/28. doi: 10.1016/j.bpj.2012.05.015. PubMed PMID: 22735543; PubMed Central PMCID: PMCPMC3379620.

Leica Microsystems. (2021). *The Guide to STED sample preparation.* Available from: https://downloads.leica-microsystems.com/STELLARIS%20STED/Application%20Notes/Article-The_Guide_to_STED_Sample_Preparation_print_MC-0002370_25032021.pdf.

Leutenegger, M., Eggeling, C., & Hell, S. W. (2010). Analytical description of STED microscopy performance. *Optics Express, 18*(25), 26417–26429. Epub 2010/12/18. https://doi.org/10.1364/OE.18.026417. PMID: 21164992.

McCann, F. E., Vanherberghen, B., Eleme, K., Carlin, L. M., Newsam, R. J., Goulding, D., et al. (2003). The size of the synaptic cleft and distinct distributions of filamentous actin, ezrin, CD43, and CD45 at activating and inhibitory human NK cell immune synapses. *Journal of Immunology, 170*(6), 2862–2870. Epub 2003/03/11. PubMed PMID: 12626536.

Menager, M. M., Menasche, G., Romao, M., Knapnougel, P., Ho, C. H., Garfa, M., et al. (2007). Secretory cytotoxic granule maturation and exocytosis require the effector protein hMunc13-4. *Nature Immunology*, *8*(3), 257–267. Epub 2007/01/24. doi: ni1431 [pii]10.1038/ni1431. PubMed PMID: 17237785.

Meyer, L., Wildanger, D., Medda, R., Punge, A., Rizzoli, S. O., Donnert, G., et al. (2008). Dual-color STED microscopy at 30-nm focal-plane resolution. *Small*, *4*(8), 1095–1100. https://doi.org/10.1002/smll.200800055 (PubMed PMID: 18671236).

Mueller, V., Ringemann, C., Honigmann, A., Schwarzmann, G., Medda, R., Leutenegger, M., et al. (2011). STED nanoscopy reveals molecular details of cholesterol- and cytoskeleton-modulated lipid interactions in living cells. *Biophysical Journal*, *101*(7), 1651–1660. https://doi.org/10.1016/j.bpj.2011.09.006. Epub 2011/10/04. PubMed PMID: 21961591; PubMed Central PMCID: PMC3183802.

Nieuwenhuizen, R. P., Lidke, K. A., Bates, M., Puig, D. L., Grunwald, D., Stallinga, S., et al. (2013). Measuring image resolution in optical nanoscopy. *Nature Methods*, *10*(6), 557–562. https://doi.org/10.1038/nmeth.2448. Epub 2013/04/30. PubMed PMID: 23624665; PubMed Central PMCID: PMCPMC4149789.

Orange, J. S., & Ballas, Z. K. (2006). Natural killer cells in human health and disease. *Clinical Immunology*, *118*(1), 1–10. Epub 2005/12/13. doi: S1521-6616(05)00334-7 [pii] 10.1016/j.clim.2005.10.011. PubMed PMID: 16337194.

Stanly, T. A., Fritzsche, M., Banerji, S., Garcia, E., de la Serna, J. B., Jackson, D. G., et al. (2016). Critical importance of appropriate fixation conditions for faithful imaging of receptor microclusters. *Biology Open*, *5*(9), 1343–1350. https://doi.org/10.1242/bio.019943. PubMed PMID: 27464671; Epub 2016/07/29. PubMed Central PMCID: PMCPMC5051640.

Stinchcombe, J. C., Bossi, G., Booth, S., & Griffiths, G. M. (2001). The immunological synapse of CTL contains a secretory domain and membrane bridges. *Immunity*, *15*(5), 751–761. Epub 2001/12/01. doi: S1074-7613(01)00234-5 [pii]. PubMed PMID: 11728337.

Wang, P., Hecht, F., Ossato, G., Tille, S., Fraser, S. E., & Junge, J. A. (2021). Complex wavelet filter improves FLIM phasors for photon starved imaging experiments. *Biomedical Optics Express*, *12*(6), 3463–3473. https://doi.org/10.1364/BOE.420953. Epub 2021/07/06. PubMed PMID: 34221672; PubMed Central PMCID: PMCPMC8221945 Microsystems that manufactures and sells the confocal microscopes (SP8 DIVE FALCON, SP8 STED FALCON, and STELLARIS 8 FALCON) used throughout the study.

> CHAPTER TWO

Imaging the immune synapse: Three-dimensional analysis of the immune synapse

Javier Ruiz-Navarro[a], Sofía Blázquez-Cucharero[a], Víctor Calvo[b], and Manuel Izquierdo[a,*]

[a]Instituto de Investigaciones Biomédicas Alberto Sols CSIC-UAM, Madrid, Spain
[b]Departamento de Bioquímica, Instituto de Investigaciones Biomédicas Alberto Sols CSIC-UAM, Facultad de Medicina, Universidad Autónoma de Madrid, Madrid, Spain
*Corresponding author: e-mail address: mizquierdo@iib.uam.es

Contents

1. Introduction	16
1.1 The immunological synapse	16
1.2 Polarization of centrosome and secretory granules	17
1.3 Synaptic actin cytoskeleton regulation of secretory traffic	18
2. Materials, cells, immunological synapse formation and image capture	20
3. 3D analysis of the IS	21
3.1 Image setup and formatting	21
3.2 ImageJ 3D synapse analysis	21
3.3 NIS-elements AR 3D synapse analysis	26
4. Discussion, future perspectives and concluding remarks	31
5. Notes	33
Conflict of interest	34
Author contributions	34
Acknowledgments	35
References	35

Abstract

T cell receptor (TCR) stimulation of T lymphocytes by antigen bound to the major histocompatibility complex (MHC) of an antigen-presenting cell (APC), together with the interaction of accessory molecules, induces the formation of the immunological synapse (IS), the convergence of secretion vesicles toward the centrosome, and the polarization of the centrosome to the IS. Upon IS formation, an initial increase in cortical filamentous actin (F-actin) at the IS takes place, followed by a decrease in F-actin density at the central region of the IS, which contains the secretory domain. These reversible, cortical actin cytoskeleton reorganization processes that characterize a mature IS occur during lytic granule secretion in cytotoxic T lymphocytes (CTL) and natural killer (NK) cells and cytokine-containing vesicle secretion in T-helper (Th) lymphocytes. Besides,

---

Methods in Cell Biology, Volume 193
ISSN 0091-679X
https://doi.org/10.1016/bs.mcb.2023.04.003

Copyright © 2025 Elsevier Inc.
All rights are reserved, including those
for text and data mining, AI training,
and similar technologies.

IS formation constitutes the basis of a signaling platform that integrates signals and coordinates molecular interactions that are necessary for an appropriate antigen-specific immune response. In this chapter we deal with the three-dimensional (3D) analysis of the synaptic interface architecture, as well as the analysis of the localization of different markers at the IS.

Abbreviations

3D	three dimensional
APC	antigen-presenting cell
BCR	B-cell receptor for antigen
CMAC	CellTracker™ Blue (7-amino-4-chloromethylcoumarin)
cSMAC	central supramolecular activation cluster
CTL	cytotoxic T lymphocytes
Dia1	diaphanous-1
dSMAC	distal supramolecular activation cluster
F-actin	filamentous actin
FCS	fetal calf serum
FMNL1	formin-like 1
IS	immunological synapse
LSFM	light sheet fluorescence microscopy
LUTs	look up tables
MFI	mean fluorescence intensity
MHC	major histocompatibility complex
MTOC	microtubule-organizing center
MVB	multivesicular bodies
NIS-AR	NIS-Elements AR
NK	natural killer
PKCδ	protein kinase C δ isoform
pSMAC	peripheral supramolecular activation cluster
ROI	region of interest
RT	room temperature
SEE	Staphylococcal enterotoxin E
SIM	structured illumination microscopy
SMAC	supramolecular activation cluster
TCR	T-cell receptor for antigen
Th	T-helper
TIRFM	total internal reflection microscopy
TRANS	transmittance

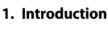

1. Introduction

1.1 The immunological synapse

The IS is a specialized cell–cell interface contact area, established between a T lymphocyte and an APC carrying cognate MHC-bound antigen, that provides a signaling platform for integration of signals leading to execution of an immune response. This allows intercellular information exchange, in order to ensure efficient TCR signal transduction, T cell activation and the proper execution of diverse T lymphocyte effector functions (Dustin & Choudhuri, 2016). IS establishment is a very dynamic, plastic and critical event, enabling the integration of spatial, mechanical and biochemical signals, involved in specific, cellular and humoral immune responses (De La Roche, Asano, & Griffiths, 2016; Fooksman et al., 2010). The general architecture of the IS upon cortical actin reorganization is recognized by the formation of a concentric, bullseye spatial pattern, named the supramolecular activation complex (SMAC) (Billadeau, Nolz, & Gomez, 2007; Carisey, Mace, Saeed, Davis, & Orange, 2018; Griffiths, Tsun, & Stinchcombe, 2010; Kuokkanen, Sustar, & Mattila, 2015; Yuseff, Pierobon, Reversat, & Lennon-Dumenil, 2013). This reorganization yields three differentiated regions: first, a central cluster of antigen receptors bound to antigen called central SMAC (cSMAC) produced by centripetal traffic; second, a surrounding ring rich in adhesion molecules, such as the LFA-1 integrin, called peripheral SMAC (pSMAC), which appears to be crucial for adhesion with the APC (Fooksman et al., 2010; Monks, Freiberg, Kupfer, Sciaky, & Kupfer, 1998); third, at the edge of the contact area with the APC, the pSMAC-surrounding distal SMAC (dSMAC), which comprises a circular array of dense F-actin (Griffiths et al., 2010; Le Floc'h & Huse, 2015; Rak, Mace, Banerjee, Svitkina, & Orange, 2011; Ritter, Angus, & Griffiths, 2013). More recently, several super-resolution imaging techniques have revealed that at least four discrete F-actin networks form and maintain the shape and function of this canonical IS upon TCR-antigen interaction (Blumenthal & Burkhardt, 2020; Hammer, Wang, Saeed, & Pedrosa, 2018). However, apart of this "classical" bullseye IS structure, non-classical IS including multifocal IS (Figs. 1 and 2) generated by double-positive thymocytes and Th2 cells, have also been observed (Brossard et al., 2005; Kumari, Colin-York, Irvine, & Fritzsche, 2019; Thauland & Parker, 2010).

1.2 Polarization of centrosome and secretory granules

IS formation induces the convergence of cellular secretion vesicles toward the centrosome and, simultaneously, the polarization of this centrosome (the major microtubule-organizing center (MTOC) in lymphocytes) toward the IS (De La Roche et al., 2016; Huse, 2012). These traffic events, acting together, lead to polarized secretion of: extracellular vesicles and exosomes coming from multivesicular bodies (MVB) by T, B lymphocytes and NK cells (Alonso et al., 2011; Calvo & Izquierdo, 2020; Herranz et al., 2019; Peters et al., 1991); lytic granules by CTL and NK cells (De La Roche et al., 2016; Ritter et al., 2013; Stinchcombe, Majorovits, Bossi, Fuller, & Griffiths, 2006); stimulatory cytokines by Th cells (Huse, Quann, & Davis, 2008); lytic proteases by B lymphocytes (Yuseff et al., 2013).

1.3 Synaptic actin cytoskeleton regulation of secretory traffic

1.3.1 Cortical actin cytoskeleton

Actin cytoskeleton plays a pivotal role from the first step of IS formation, coordinating its assembly and driving, together with microtubules, the accumulation of synaptic components that are required for T cell activation (Ritter et al., 2013). However, the involvement of F-actin in IS extends beyond the rearrangement and signaling events that take place at the first steps, since F-actin also plays a crucial role in IS maintenance and in antigen receptor-derived signaling in T and B lymphocytes, ensuring a functional immune response (Billadeau et al., 2007). Refer to the excellent reviews on the IS formed by B lymphocytes (Yuseff et al., 2013; Yuseff, Lankar, & Lennon-Dumenil, 2009) and T lymphocytes (Billadeau et al., 2007; Ritter et al., 2013).

At the early phases of IS formation, a protrusive actin polymerization activity drives the radially symmetric spreading of the T cell over the surface of the APC through the generation of filopodia and lamellipodia (Le Floc'h & Huse, 2015). As the IS evolves, cortical F-actin density progressively decreases in the cSMAC, while F-actin reorganizes and accumulates into the dSMAC, creating a peripheral F-actin-rich ring, and allowing the directional secretion toward the APC by congregating secretion vesicles on the IS (Stinchcombe et al., 2006). Thus, F-actin forms a permissive network at the IS of CTL and NK cells (Carisey et al., 2018; Ritter et al., 2015).

The concentric F-actin architecture and the cortical actin cytoskeleton reorganization are shared by IS formed by B lymphocytes, CD4+ Th lymphocytes, CD8+ CTL, and NK cells (Brown et al., 2011; Le Floc'h & Huse, 2015). Remarkably, all these immune cells form IS and directionally secrete proteases, cytokines or cytotoxic factors at the IS. This directional secretion, regulated by F-actin synaptic architecture, most probably enhances the specificity and the efficacy of the subsequent responses to these molecules (Le Floc'h & Huse, 2015), by spatially and temporally focusing the secretion at the synaptic cleft (Billadeau et al., 2007), which avoids the stimulation or death of bystander cells. However, in this paper we deal only with IS made by T lymphocytes, although the method can be extended to the IS made by the other immune cells.

1.3.2 Cortical and non-cortical actin cytoskeleton regulation of secretory traffic

F-actin reduction at the cSMAC does not just allow secretion, since it apparently also plays a key role in MTOC movement toward the IS (Ritter et al., 2015; Stinchcombe et al., 2006). IS-induced actin cytoskeleton reorganization and secretory vesicles polarized traffic are regulated by two major pathways: one involves HS1/WASp/Arp2/3 complexes acting on cortical F-actin, and the other involves formins, a conserved family of proteins, with members such as formin-like 1 (FMNL1) and Diaphanous 1 (Dia1), that nucleate and elongate unbranched actin filaments (Deward, Eisenmann, Matheson, & Alberts, 2010; Kühn & Geyer, 2014; Kumari, Curado, Mayya, & Dustin, 2014). In this context, several studies support, using 3D analysis of the IS with high spatio-temporal resolution, that cortical F-actin reorganization involving an initial F-actin accumulation at the cell-to-cell contact area, and subsequent F-actin clearing at the central IS, is necessary and sufficient for centrosome and secretory granule polarization (Chemin et al., 2012; Ritter et al., 2015; Sanchez, Liu, & Huse, 2019). However, other results show that FMNL1 or Dia1 depletion impedes centrosome polarization without affecting Arp2/3-dependent cortical F-actin accumulation (Gomez et al., 2007) supporting that, at least in the absence of FMNL1 or Dia1, initial cortical F-actin accumulation is not sufficient for centrosome polarization. However, in the last report neither 3D nor kinetic analysis of the sub-synaptic architecture was performed, thus no studies to assess F-actin clearing at the central IS were performed. Conversely, the centrosome can polarize normally to the IS in the absence of cortical actin accumulation at the IS occurring in Jurkat T lymphocytes lacking Arp2/3 (Gomez et al., 2007; Kumari et al., 2014). Thus, analysis of all the F-actin networks at different subcellular locations and at diverse sub-synaptic regions by 3D analyses, but also at different time points, upon IS formation is necessary to achieve the full picture of the cellular actin cytoskeleton reorganization processes leading to polarized secretion. These results, together with our results showing that PKCδ interference affects cortical F-actin at the IS and F-actin depletion at the central IS (Herranz et al., 2019), and PKCδ regulates the phosphorylation of FMNL1 (Bello-Gamboa et al., 2020), prompted us to study the involvement and sub-synaptic distribution of cortical F-actin and FMNL1 formin in T lymphocyte IS structure. The experimental approach described here allows the simultaneous assessment of the sub-synaptic architecture and colocalization of two different markers in the three dimensions of the IS. This method is properly developed for confocal, but also for super-resolution microscopy images and can be used to assess cortical F-actin in different types of IS, including the synapses made by NK cells and B lymphocytes, since these synapses share similar organization to those developed by T lymphocytes (Calvo & Izquierdo, 2021; Ritter et al., 2013; Yuseff et al., 2013, 2009). As already mentioned, the crucial importance of the synaptic structure in processes such as centrosome polarization or secretory vesicles traffic and degranulation has stimulated the development of accurate analysis methods of the synaptic interface. Moreover, considering that actual biological interactions occur in 3D space, following this approach, we can study the architecture of the IS and the 3D colocalization and potential interactions between different markers in this region, thus overcoming the limitations of two-dimensional analyses.

2. Materials, cells, immunological synapse formation and image capture

1. Raji B cell line is obtained from the ATCC and Jurkat T clone C3 has been described (Herranz et al., 2019).
2. Cell lines are cultured in RPMI 1640 medium supplemented with L-glutamine (Invitrogen), 10% heat-inactivated fetal calf serum (FCS) (Gibco) and penicillin/streptomycin (Gibco) and 10 mM HEPES (Lonza).
3. For IS formation, Raji cells are attached to μ-Slide 8 Well Glass Bottom (Ibidi), using 20 μg/mL poly-L-lysine (SIGMA) at 37 °C for 1 h, and then labeled with 10 μM CellTracker™ Blue (7-amino-4-chloromethylcoumarin, CMAC, ThermoFisher), in supplemented RPMI 1640 medium at 37 °C for 45 min, and pulsed with 1 μg/mL Staphylococcal enterotoxin E (SEE, Toxin Technology, Inc.) at 37 °C for 45 min. After removing the culture medium containing SEE, Jurkat cells are directly added to the wells, and incubated at 37 °C for 1 h, so that IS are formed. CMAC labeling allows discriminating

Jurkat-Raji conjugates (Figs. 1 and 2). Please refer to the following references for further details, since this IS model has been exhaustively described (Bello-Gamboa et al., 2019; Montoya et al., 2002). Endpoint fixation is performed with 4% paraformaldehyde at room temperature (RT) for 15 min, and subsequently with chilled acetone for 5 min (see *Note 1*).
4. Immunofluorescence is performed as described (Fernández-Hermira, Sanz-Fernández, Botas, Calvo, & Izquierdo, 2023). FMNL1 is labeled with mouse monoclonal anti-FMNL1 (clone C5, Santa Cruz Biotechnology) at RT for 45 min, and an appropriate secondary antibody coupled to AF546 (ThermoFisher) at RT for 30 min. F-actin is labeled with phalloidin AF647 (ThermoFisher) at RT for 30 min. CMAC fluorescence does not overlap with AF546 or AF647 fluorescence.
5. Image capture is performed with Leica SP8 confocal microscope. Around 30 optical sections are acquired using a 0.2–0.3 μm Z-step size to yield an appropriate axial resolution. Confocal settings are same as those previously used (Fernández-Hermira et al., 2023).

3. 3D analysis of the IS

3.1 Image setup and formatting

1. Open the confocal file in the public, multiplatform software ImageJ or FIJI (https://imagej.nih.gov/ij/, https://fiji.sc), by selecting "Open" > "Select File" > "OK." All channels saved for each selected image should open automatically using "Bioformats import" plugin. Depending on the confocal file source it could be necessary to specify pixel width, height and voxel depth in microns. Please include in "Image" > "Properties" these parameters obtained from confocal image metadata. This also allows proper image calibration.
2. Save all individual channels in TIFF format by using "File" > "Save As" > "OK."

3.2 ImageJ 3D synapse analysis

1. Channel rotation
   1.1 Open a TIFF file in ImageJ or FIJI by selecting "Open" > "Select File" > "OK." All channels saved for each selected image should open automatically.

1.2 Rotate channel by using "Image" > "Transform" > "Rotate." Enter desired angle so as to position the Raji B cell vertically to the Jurkat T cell. The synapse should appear horizontally to the observer. "Preview" function allows to view the image with the selected angle. By clicking "OK," modifications are saved. Repeat process with the same angle for all channels.

1.3 Save all individual channels by using "File" > "Save As" > "Tiff" > "OK."

2. Channel crop

2.1 To facilitate cropping, synchronize all channels by using "Analyze" > "Tools" > "Synchronize Windows." Select "Synchronize All" in pop-up window. Channels stay synchronized between each other as long as the pop-up window stays open.

2.2 Select area to crop using the "rectangle" tool. Same area and same position are reproduced in all synchronized crops. Use "Image" > "Crop." Select an area comprising both the entire Raji B and Jurkat T cells (Fig. 1A). A second crop can be generated in order to select only the synapse area between the two cells (Fig. 1A, white rectangle and Fig. 1B). If channels are not synchronized, setting up the same width "w" and height "h" values, and "x" and "y" positions, ensures the equivalence in size and position between all cropped channels. Values for "w," "h," "x" and "y" are displayed in FIJI's console. Crop deserves special attention, since it is of great importance to know the area in which the synapse occurs in order to be able to perform a precise crop (see *Note 2*). Since these are studies of "real" immune synapses, there is a great variety of morphologies among the synapses studied, so the regions of contact may vary from one to another. The cortical F-actin structures can be usually distinguished in the Jurkat cells. In other cases, in which these structures are not easily identified, the bright field images of the interacting cells can be of great assistance to this matter. It is preferable to select a broader, rectangular region of interest (ROI) including all the synapse area instead an irregular, very restricted ROI, since this can exclude in Z dimension relevant interleaving areas between the cells. Finally, the use of markers or cell trackers specific of one of both cells, such as CMAC, may help to localize the contact zone.

2.3 Save rotated cropped channels as TIFF files.

3. Generation of the 3D file and video recording

3.1 Image look up tables (LUTs) can be adjusted by using "Image" > "Adjust" > "Brightness/Contrast."

3.2 Merge two or more channels by using "Image" > "Color" > "Merge channels." Choose color components for each channel and select "OK" (see Note 3). Select "Keep source images" to keep original images.

3.3 Create a 3D image by selecting "Plugins" > "3D Viewer." In "Image," select desired file for video. Leave other parameters as given. Click "OK."

3.4 To visualize the delimitating box for the video, select "Edit" > "Show bounding box."

3.5 To generate video, select "View" > "Start freehand recording" and create sequential frames by using the arrows in the keyboard (see Note 4). Alternatively, select "Set View" to choose the desired plane.

3.6 To stop recording, select "View" > "Stop freehand recording." A window named "Movie" appears.

3.7 "Movie" file can be saved by using "File" > "Save As" > "AVI." Once the .avi stack is generated (Supplementary Video 1 in the online version at https://doi.org/10.1016/bs.mcb.2023.04.003) the video stack can be converted to sequential images by using "Image" > "Stacks" > "Stack to images" (Supplementary Video 1 in the online version at https://doi.org/10.1016/bs.mcb.2023.04. 003 and Fig. 1B).

4. Quantification and measurement of IS interface (en face) intensity profile

4.1 In order to obtain a synapse's intensity profile for each channel, individual recordings should be generated for all channels included in the video.

4.2 Scroll aforementioned videos to those frames where the interface of the synapsis is shown perpendicular (en face) to the observer.

4.3 Select area over the en face frame to create the intensity profile. Save this by using "Analyze" > "Tools" > "ROI Manager." Click on "Add" and "More" > "Save."

4.4 To obtain the intensity profile graph, select "Analyze" > "Plot Profile." Save each individual channel graph by using "File" > "Save As" > "Tiff."

4.5 To obtain numerical data, select "List" in graph and copy the data onto an Excel spreadsheet. By copying the individual data from all channels and inserting the desired graph, a new intensity profile plot can be generated that emulates all chosen channels (Fig. 1C).

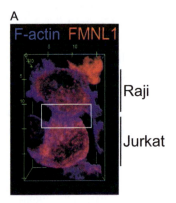

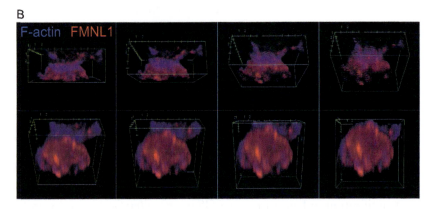

Fig. 1 ImageJ-based 3D analyses of the immunological synapse. Jurkat cells were challenged with CMAC-labeled (blue), SEE-pulsed Raji cells for 1 h, then they were fixed, stained with anti-FMNL1 AF546 to label FMNL1 (red) and phalloidin AF647 to label F-actin (magenta) and imaged by confocal fluorescence microscopy. F-actin (magenta) was changed to blue for 3D visualization. (A) Top view of a whole synaptic conjugate labeled with anti-FMNL1 (red) and F-actin (blue). The white rectangle labels the crop that contains the IS area analyzed in (B). (B) Sequential frames (one each 20 frames) taken from Supplementary Video 1 in the online version at https://doi.org/10.1016/bs.mcb.2023.04.003. (C) Last frames from Supplementary Video 1 in the online version at https://doi.org/10.1016/bs.mcb.2023.04.003 (IS interface) and corresponding mean fluorescence intensity (MFI) intensity plots of FMNL1 (red) and F-actin (blue) channels. F-actin profile reveals a multifocal IS, with several low F-actin MFI areas. The FMNL1 profile (red line) is partially similar to F-actin profile.

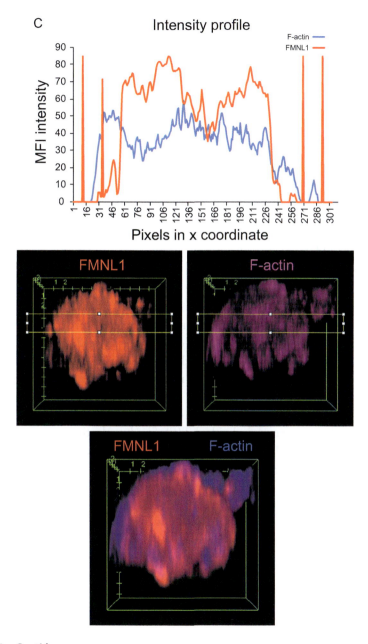

Fig. 1—Cont'd

3.3 NIS-elements AR 3D synapse analysis

1. Channel rotation

   1.1 Open the software NIS-Elements AR (https://www.microscope. healthcare.nikon.com/en_EU/products/software/nis-elements/nis-elements-advanced-research) (NIS-AR) and a TIFF confocal file generated in Section 3.1 by using "Image" > "Open" > "OK."

   1.2 Rotate channel by using "Image" > "Rotate" > "Rotate by Degree." Enter desired angle so as to position the Raji B cell vertically to the Jurkat T cell. The synapse should appear horizontally to the observer.

   1.3 Repeat process with the same angle rotation for all channels. Save rotated channels as separated, individual TIFF files by selecting "File" > "Save/Export to TIFF Files" > "Choose Output Folder" > "Export."

2. Channel crop

   2.1 Open rotated TIFF channels and select "Image" > "Crop." Select area comprising both the entire Raji B and Jurkat T cells (Fig. 2A). A second crop can be generated in order to show only the synapse between the two cells (see *Note 2*) (Fig. 2A, white rectangle and Fig. 2B). By ticking "Remember Last Settings" tool, all subsequent crops should measure the same and be located at the same x, y position.

   2.2 Repeat process for all rotated channels. Save rotated cropped channels as separated, individual TIFF files by selecting "File" > "Save/Export to TIFF Files" > "Choose Output Folder" > "Export."

3. Generation of the 3D file

   3.1 Open two equally edited channel TIFF files by using "File" > "Open" > "OK."

   3.2 The "Experiment Type" window appears automatically. Enter Z series width for each channel.

   3.3 Merge both channels by using "File" > "Merge Channels." Choose RGB color components for each channel (see *Note 3*).

   3.4 Image LUTs can be adjusted by using "Visualization Controls" > "LUTs."

   3.5 Colocalization analysis
   - Show fluorocytogram by selecting "View" > "Analysis Controls" > "Colocalization." Define colocalization region with "Rectangular Selection" or "Sector Selection" symbols (see *Note 5*).

- Change colocalization color to white, "View" > "Analysis Controls" > "Binary Layers" > "Colocalization" > "More" > "Select desired color" > "Accept," to distinguish better the colocalization pixels from the other channel colors.
- Save the colocalization selected region by selecting "Save/Load Colocalize Configuration" symbol > "Save Colocalization Settings" > "Save."

3.6 Create a 3D image by selecting "Show Volume View" symbol on the top toolbar (Fig. 2B).

4. Video recording of the 3D file

4.1 While on the 3D generated image, open the video editor with "Show Movie Maker" tool on the top toolbar.

4.2 Select the desired starting frame by using manual tools or by selecting "View Plane" on the top toolbar and choosing the desired plane. Select "Plus" symbol on the bottom of the screen to fix initial settings.

4.3 Repeat action as many times as needed to create the final movie. Added blocks can be individually edited by right-clicking on them.

4.4 Select "Play/Stop Movie" to show the recording.

4.5 Save selected recording by using "Export Movie" symbol on the bottom of the screen. Save the generated recording by clicking "File" > "Save As" both in .nd2 and .avi formats for future editing and viewing. Once the .avi stack is generated (Supplementary Video 2 in the online version at https://doi.org/10.1016/bs.mcb.2023.04.003) the video stack can be converted to serial images by using "Image" > "Stacks" > "Stack to images" in ImageJ (Fig. 2B).

5. Quantification and measurement of synapse's en face intensity profile

5.1 Display the intensity profile tool in NIS-AR.

5.2 Open the desired video in .nd2 format by using "File" > "Open."

5.3 Scroll the video to the time frame where the IS interface is shown perpendicular to the observer (en face) by reproducing the video through all its time frames using the bottom blue T level bar.

5.4 Select the "Intensity Profile" symbol on the top toolbar and adjust the Profile Line by dragging it until it is positioned in the desired area (to analyze colocalization data, see *Notes 6* and *7*). The x, y coordinates of the Profile Line can be viewed at the bottom of the screen, and used, if necessary, in other videos.

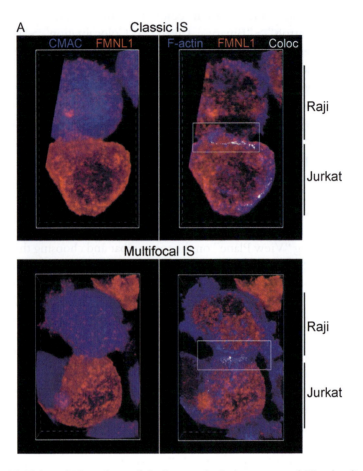

Fig. 2 NIS-AR-based 3D analyses of the immunological synapse and 3D colocalization. Jurkat cells were challenged with CMAC-labeled (blue), SEE-pulsed Raji cells for 1 h, then they were fixed, stained with anti-FMNL1 AF546 to label FMNL1 (red) and phalloidin AF647 to label F-actin (magenta) and imaged by confocal fluorescence microscopy. F-actin (magenta) was changed to blue for 3D co-localization as required in NIS-AR. In (A), upper panel shows the top view of a whole "classic" synaptic conjugate (Raji cell labeled with CMAC, blue; anti-FMNL1, red), whereas lower panel displays the top view of a multifocal synaptic conjugate, showing 3D colocalization pixels between FMNL1 (red) and F-actin (blue) in white color in the right side panels. The white rectangle labels the crop area that contains the IS area analyzed in (B).

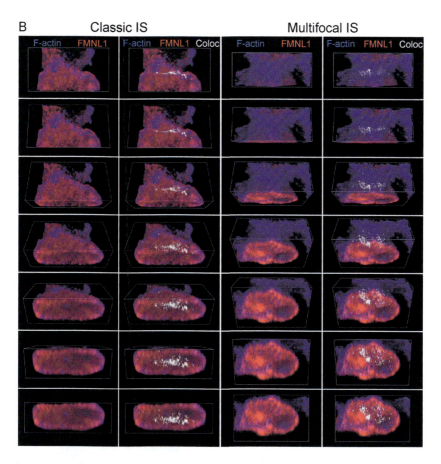

Fig. 2—Cont'd (B) Sequential frames (one each 20 frames) taken from Supplementary Video 2 in the online version at https://doi.org/10.1016/bs.mcb.2023.04.003, showing 3D colocalization pixels (white color) on each frame. Colocalization pixels between FMNL1 and F-actin accumulate at the central area of the IS in both examples.

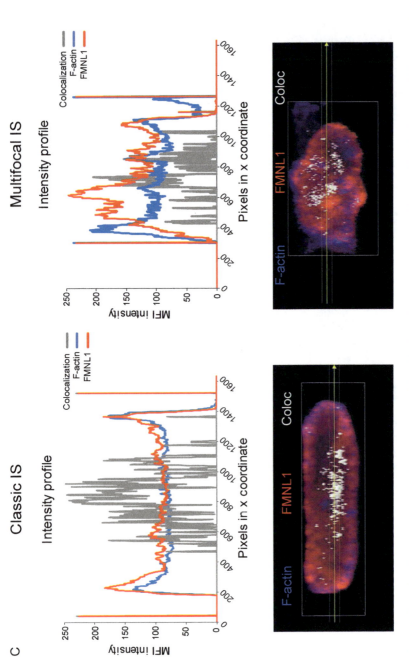

Fig. 2—Cont'd (C) Last frame from Supplementary Video 2 in the online version at https://doi.org/10.1016/bs.mcb.2023.04.003 (IS interface) and MFI intensity plots of FMNL1 (red), F-actin (blue) channels and colocalization (gray). F-actin profile reveals a classic (containing a wide, single F-actin low area at the central IS; left) and a multifocal IS, with several F-actin low areas (right). In general, F-actin density (blue line) at the central IS, that contains the cSMAC, is lower than at the edges (corresponding to the dSMAC). FMNL1 profile is partially similar to F-actin profile, decreasing at the center of the IS, whereas the F-actin and FMNL1 colocalization (gray line) is higher at the central IS.

5.5 Upon generation of the intensity profile graph, separate profiles of each selected channels and the colocalization channel are shown in the "Intensity Profile" window.

5.6 Export the data to an Excel spreadsheet by selecting "Export" on the "Intensity Profile" window. A new Excel document is automatically generated with the raw data. By selecting the appropriate data and inserting the desired graph, a new separate Intensity Profile graph for each channel, including the colocalization channel, can be generated (Fig. 2C).

4. Discussion, future perspectives and concluding remarks

We have noticed that the quality of the 3D videos generated by NIS-AR is higher than that of videos generated using ImageJ/3D viewer, and that both lagging and poor resolution can be appreciated in some frames of ImageJ-generated videos (compare Figs. 1 and 2). In addition, NIS-AR software easily handles the files, whereas we have observed certain software instability during ImageJ processing in three different computers. In addition, 3D colocalization plugin is not available in ImageJ yet, please check for plugin actualizations in ImageJ website (https://imagej.nih.gov/ij/). It is important to remark that the procedure described here is intended to handle post-processing of confocal imaging in fixed samples, but there are some alternatives for 3D reconstitution and IS interface visualization of the IS in living cells with high temporal resolution, albeit the equipment involved is scarce and expensive (Calvo & Izquierdo, 2018). These techniques allow image analyses of developing synapses carried out during the acquisition stage, by using light sheet fluorescence microscopy (LSFM) (Calvo & Izquierdo, 2018; Ritter et al., 2015). Indeed, LSFM technique allows four-dimensional (x, y, z, t) analyses of IS interface and F-actin architecture using SiR-Actin probes in living cells, but the superresolution option is not available at this moment using this technique. In addition, other approaches based on stimulatory artificial surfaces (lipid bilayers) and total internal reflection microscopy (TIRFM), combined with superresolution techniques such as structured illumination microscopy (SIM) allow superresolution images of IS interface and superb F-actin imaging in living synapses, by reducing 3D dimensions to two (Calvo & Izquierdo, 2018; Carisey et al., 2018; Murugesan et al., 2016), although these approaches are intrinsically reductionist and do not mimic the complex interactions and the irregular stimulatory surface taking place in a "real" cell–cell synapse (Dustin, 2009) as we present here. It is important to remark that MHC-II-antigen triggering on the B cell side of the Th synapse does not induce noticeable F-actin changes (Calvo & Izquierdo, 2021; Yuseff et al., 2013). In addition, we have observed (i.e., Fig. 1) that majority of F-actin at the IS belongs to the Jurkat cell. Thus, it is expected that the contribution to the analyses of the residual, invariant F-actin from the B cell will be negligible using this protocol. In addition, the 3D analyses described here can indeed distinguish multifocal from "classical" IS (compare the two examples in Fig. 2).

As described, the protocols can be used to handle files generated in diverse confocal microscopes due to the flexibility of ImageJ. Thus, it is conceivable that the resolution and quality of the generated videos will greatly improve with the enhanced resolution and quality of the source images, including superresolution images (Calvo & Izquierdo, 2018).

The method described above is not automated and to perform the quantification the user must therefore select single cells forming IS one by one, which may make it difficult to evaluate results in a fully unbiased manner and may also be time-consuming to execute for an untrained user. However, a trained user may analyze up to 4–5 synapses per hour. To prevent biases, please try to randomly select as many synaptic conjugates from different fields as possible. Unfortunately, it is not easy to distinguish canonic IS from irrelevant cell contacts or cell aggregates randomly produced by the high cell concentrations used to favor IS formation (Bello-Gamboa et al., 2019). Evaluation of the cup-shaped profile of the effector T lymphocyte and/or lamellipodium formation observed in the transmittance channel may help to identify IS (Fernández-Hermira et al., 2023). In addition, a useful criterion to unambiguously select canonic IS and differentiate them from irrelevant cell contacts is to use the F-actin probe phalloidin (Figs. 1 and 2), since it is known that F-actin accumulates at the lamellipodium in the synaptic contact area (see paragraph 1. Introduction). CMAC labeling facilitates discrimination of Jurkat-Raji conjugates from Jurkat-Jurkat or Raji-Raji homotypic conjugates.

In addition, and due to the fact that the IS characteristic concentric bullseye F-actin architecture and the actin cortical cytoskeleton reorganization processes are shared by IS formed by B lymphocytes, CD4+ Th lymphocytes, CD8+ CTL, and NK cells forming IS (Brown et al., 2011; Le Floc'h & Huse, 2015), the protocol described here can be extended to analyze IS formed by all these types of immune cells. This analysis will enable to address how the different cortical F-actin network areas in all these IS classes are regulated, how these networks integrate into cell surface receptor-evoked signaling networks, as well as their potential regulators, that constitute an intriguing and challenging biological issue (Calvo & Izquierdo, 2021; Dogterom & Koenderink, 2019).

5. Notes

1. Acetone (and acetone vapor) dissolves plastic, thus keep the μ-Slide with fixative on ice at 4 °C and remove the μ-Slide plastic lid. Recommended maximal fixation time is 5 min, since longer times dehydrate samples in excess (producing cell shrinking and cell shape changes). Paraformaldehyde or glutaraldehyde fixatives alone do not allow a clear staining of FMNL1. An important issue is that F-actin staining with phalloidin is not compatible with methanol fixing, whereas it is compatible with paraformaldehyde and/or acetone fixation. Thus, we have chosen paraformaldehyde/acetone fixation since it constitutes a good compromise to circumvent these caveats (Abrahamsen, Sundvold-Gjerstad, Habtamu, Bogen, & Spurkland, 2018).

2. It is important to check all Z optical sections so as to avoid cropping out an area that includes part of one or both cells. Cropped areas can be adjusted to include the entirety of the cell area at different Z optical sections. The crops of all channels of each image must be equal both in size and x, y position.

3. For F-actin and FMNL1 staining experiment, Blue and Red Components were chosen, respectively. Please realize that 3D co-localization in NIS-AR only works with Red, Green and Blue channels, therefore F-actin (acquired in magenta) was changed to blue color for this purpose (see Figs. 1 and 2).

4. The speed at which the keyboard arrows are clicked does not affect the final speed at which the video is recorded.

5. Once the colocalization settings are defined, load them for all Zs by scrolling the image stack through all its Z optical sections, using either the scroll wheel on the mouse or dragging them from the bottom blue Z level bar. If not, it is possible that in the subsequent 3D visualization the updated colocalization data will not be loaded.

6. Due to the properties of the video editor and the three-color nature of the "Merge Channels" tool, the generated videos will by default be RGB. For that reason, caution should be taken on the intensity profile

measurements made with colocalization videos as, depending on the selected colocalization color, most probably some color components will not be properly separated in the intensity profile. For example, if, as suggested above, colocalization is displayed in white, every signal from our markers will be contaminated by the colocalization measurement. Moreover, the different relative intensities between studied channels, being in the same merge document, may affect each other visualization because one channel masking the signal of the other. This might lead to some discrepancies in the intensity profile quantification of each individual channel. In order to solve these problems, a new video with the same image of each individual channel should be created. Thus we will have one video per channel and an additional one with both channels and the colocalization data. The three different videos can be plotted in the same intensity profile graph by selecting the specific color components: for the video with colocalization the color component will be the one associated with colocalization, whereas for the videos of each channel the color components will be those colors that were selected for them when the merge was done.

7. On the Intensity Profile graph, the signal region is flanked by two intense and very well-defined lines (Fig. 2C). These lines correspond to the boundaries of the "box" containing either the whole cell conjugate or the synapse area in the 3D display. To properly compare the Intensity Profile of different 3D videos of the same cell—as in the colocalization case—the profile line should be positioned at the same x, y coordinates in every video and the width of the generated profiles should be the same. If the 3D videos display different widths, a careful look into the sizes of the two created videos should be taken.

Conflict of interest

The authors declare that the research was conducted in the absence of any commercial or financial relationships that could be construed as a potential conflict of interest.

Author contributions

J.R. and S.B. wrote the manuscript and prepared the figs. V.C. and M.I. conceived the manuscript and the writing of the manuscript and approved its final content. Conceptualization, V.C. and M.I; writing original draft preparation, M.I.; reviewing and editing, V.C. and M.I.

Acknowledgments

This research was funded by grants from the Spanish Ministerio de Ciencia e Innovación (PID2020-114148RB-I00) to MI, which was in part granted with FEDER funding (EC), corresponding to Programa Estatal de Generación de Conocimiento y Fortalecimiento Científico y Tecnológico del Sistema de i+d+i y de i+d+i Orientada a los Retos de la Sociedad.

References

Abrahamsen, G., Sundvold-Gjerstad, V., Habtamu, M., Bogen, B., & Spurkland, A. (2018). Polarity of CD4+ T cells towards the antigen presenting cell is regulated by the Lck adapter TSAd. *Scientific Reports*, *8*, 13319.

Alonso, R., Mazzeo, C., Rodriguez, M. C., Marsh, M., Fraile-Ramos, A., Calvo, V., et al. (2011). Diacylglycerol kinase alpha regulates the formation and polarisation of mature multivesicular bodies involved in the secretion of Fas ligand-containing exosomes in T lymphocytes. *Cell Death and Differentiation*, *18*, 1161–1173.

Bello-Gamboa, A., Izquierdo, J. M., Velasco, M., Moreno, S., Garrido, A., Meyers, L., et al. (2019). Imaging the human immunological synapse. *Journal of Visualized Experiments*, (154). https://doi.org/10.3791/60312.

Bello-Gamboa, A., Velasco, M., Moreno, S., Herranz, G., Ilie, R., Huetos, S., et al. (2020). Actin reorganization at the centrosomal area and the immune synapse regulates polarized secretory traffic of multivesicular bodies in T lymphocytes. *Journal of Extracellular Vesicles*, *9*, 1759926.

Billadeau, D. D., Nolz, J. C., & Gomez, T. S. (2007). Regulation of T-cell activation by the cytoskeleton. *Nature Reviews. Immunology*, 7, 131–143.

Blumenthal, D., & Burkhardt, J. K. (2020). Multiple actin networks coordinate mechanotransduction at the immunological synapse. *The Journal of Cell Biology*, *219*(2), e201911058. https://doi.org/10.1083/jcb.201911058.

Brossard, C., Feuillet, V., Schmitt, A., Randriamampita, C., Romao, M., Raposo, G., et al. (2005). Multifocal structure of the T cell—Dendritic cell synapse. *European Journal of Immunology*, *35*, 1741–1753.

Brown, A. C., Oddos, S., Dobbie, I. M., Alakoskela, J. M., Parton, R. M., Eissmann, P., et al. (2011). Remodelling of cortical actin where lytic granules dock at natural killer cell immune synapses revealed by super-resolution microscopy. *PLoS Biology*, *9*, e1001152.

Calvo, V., & Izquierdo, M. (2018). Imaging polarized secretory traffic at the immune synapse in living T lymphocytes. *Frontiers in Immunology*, *9*, 684.

Calvo, V., & Izquierdo, M. (2020). Inducible polarized secretion of exosomes in T and B lymphocytes. *International Journal of Molecular Sciences*, *21*(7), 2631.

Calvo, V., & Izquierdo, M. (2021). Role of actin cytoskeleton reorganization in polarized secretory traffic at the immunological synapse. *Frontiers in Cell and Development Biology*, *9*, 629097.

Carisey, A. F., Mace, E. M., Saeed, M. B., Davis, D. M., & Orange, J. S. (2018). Nanoscale dynamism of actin enables secretory function in cytolytic cells. *Current Biology*, *28*, 489–502. e489.

Chemin, K., Bohineust, A., Dogniaux, S., Tourret, M., Guegan, S., Miro, F., et al. (2012). Cytokine secretion by CD4+ T cells at the immunological synapse requires Cdc42-dependent local actin remodeling but not microtubule organizing center polarity. *Journal of Immunology*, *189*, 2159–2168.

De La Roche, M., Asano, Y., & Griffiths, G. M. (2016). Origins of the cytolytic synapse. *Nature Reviews. Immunology*, *16*, 421–432.

Deward, A. D., Eisenmann, K. M., Matheson, S. F., & Alberts, A. S. (2010). The role of formins in human disease. *Biochimica et Biophysica Acta, 1803*, 226–233.

Dogterom, M., & Koenderink, G. H. (2019). Actin–microtubule crosstalk in cell biology. *Nature Reviews Molecular Cell Biology, 20*, 38–54.

Dustin, M. L. (2009). Supported bilayers at the vanguard of immune cell activation studies. *Journal of Structural Biology, 168*, 152–160.

Dustin, M. L., & Choudhuri, K. (2016). Signaling and polarized communication across the T cell immunological synapse. *Annual Review of Cell and Developmental Biology, 32*, 303–325.

Fernández-Hermira, S., Sanz-Fernández, I., Botas, M., Calvo, V., & Izquierdo, M. (2023). Analysis of centrosomal area actin reorganization and centrosome polarization upon lymphocyte activation at the immunological synapse. *Methods in Cell Biology, 173*, 15–32.

Fooksman, D. R., Vardhana, S., Vasiliver-Shamis, G., Liese, J., Blair, D. A., Waite, J., et al. (2010). Functional anatomy of T cell activation and synapse formation. *Annual Review of Immunology, 28*, 79–105.

Gomez, T. S., Kumar, K., Medeiros, R. B., Shimizu, Y., Leibson, P. J., & Billadeau, D. D. (2007). Formins regulate the actin-related protein 2/3 complex-independent polarization of the centrosome to the immunological synapse. *Immunity, 26*, 177–190.

Griffiths, G. M., Tsun, A., & Stinchcombe, J. C. (2010). The immunological synapse: A focal point for endocytosis and exocytosis. *The Journal of Cell Biology, 189*, 399–406.

Hammer, J. A., Wang, J. C., Saeed, M., & Pedrosa, A. T. (2018). Origin, organization, dynamics, and function of actin and actomyosin networks at the T cell immunological synapse. *Annual Review of Immunology, 37*, 201–224.

Herranz, G., Aguilera, P., Davila, S., Sanchez, A., Stancu, B., Gomez, J., et al. (2019). Protein kinase C delta regulates the depletion of actin at the immunological synapse required for polarized exosome secretion by T cells. *Frontiers in Immunology, 10*, 851.

Huse, M. (2012). Microtubule-organizing center polarity and the immunological synapse: Protein kinase C and beyond. *Frontiers in Immunology, 3*, 235.

Huse, M., Quann, E. J., & Davis, M. M. (2008). Shouts, whispers and the kiss of death: Directional secretion in T cells. *Nature Immunology, 9*, 1105–1111.

Kühn, S., & Geyer, M. (2014). Formins as effector proteins of Rho GTPases. *Small GTPases, 5*, e29513.

Kumari, S., Colin-York, H., Irvine, D. J., & Fritzsche, M. (2019). Not all T cell synapses are built the same way. *Trends in Immunology, 40*, 977–980.

Kumari, S., Curado, S., Mayya, V., & Dustin, M. L. (2014). T cell antigen receptor activation and actin cytoskeleton remodeling. *Biochimica et Biophysica Acta, 1838*. https://doi.org/10.1016/j.bbamem.2013.1005.1004.

Kuokkanen, E., Sustar, V., & Mattila, P. K. (2015). Molecular control of B cell activation and immunological synapse formation. *Traffic, 16*, 311–326.

Le Floc'h, A., & Huse, M. (2015). Molecular mechanisms and functional implications of polarized actin remodeling at the T cell immunological synapse. *Cellular and Molecular Life Sciences, 72*, 537–556.

Monks, C. R., Freiberg, B. A., Kupfer, H., Sciaky, N., & Kupfer, A. (1998). Three-dimensional segregation of supramolecular activation clusters in T cells. *Nature, 395*, 82–86.

Montoya, M. C., Sancho, D., Bonello, G., Collette, Y., Langlet, C., He, H. T., et al. (2002). Role of ICAM-3 in the initial interaction of T lymphocytes and APCs. *Nature Immunology, 3*, 159–168.

Murugesan, S., Hong, J., Yi, J., Li, D., Beach, J. R., Shao, L., et al. (2016). Formin-generated actomyosin arcs propel T cell receptor microcluster movement at the immune synapse. *The Journal of Cell Biology, 215*, 383–399.

Peters, P. J., Borst, J., Oorschot, V., Fukuda, M., Krahenbuhl, O., Tschopp, J., et al. (1991). Cytotoxic T lymphocyte granules are secretory lysosomes, containing both perforin and granzymes. *The Journal of Experimental Medicine, 173*, 1099–1109.

Rak, G. D., Mace, E. M., Banerjee, P. P., Svitkina, T., & Orange, J. S. (2011). Natural killer cell lytic granule secretion occurs through a pervasive actin network at the immune synapse. *PLoS Biology, 9*, e1001151.

Ritter, A. T., Angus, K. L., & Griffiths, G. M. (2013). The role of the cytoskeleton at the immunological synapse. *Immunological Reviews, 256*, 107–117.

Ritter, A. T., Asano, Y., Stinchcombe, J. C., Dieckmann, N. M., Chen, B. C., Gawden-Bone, C., et al. (2015). Actin depletion initiates events leading to granule secretion at the immunological synapse. *Immunity, 42*, 864–876.

Sanchez, E., Liu, X., & Huse, M. (2019). Actin clearance promotes polarized dynein accumulation at the immunological synapse. *PLoS One, 14*, e0210377.

Stinchcombe, J. C., Majorovits, E., Bossi, G., Fuller, S., & Griffiths, G. M. (2006). Centrosome polarization delivers secretory granules to the immunological synapse. *Nature, 443*, 462–465.

Thauland, T. J., & Parker, D. C. (2010). Diversity in immunological synapse structure. *Immunology, 131*, 466–472.

Yuseff, M. I., Lankar, D., & Lennon-Dumenil, A. M. (2009). Dynamics of membrane trafficking downstream of B and T cell receptor engagement: Impact on immune synapses. *Traffic, 10*, 629–636.

Yuseff, M. I., Pierobon, P., Reversat, A., & Lennon-Dumenil, A. M. (2013). How B cells capture, process and present antigens: A crucial role for cell polarity. *Nature Reviews. Immunology, 13*, 475–486.

CHAPTER THREE

Image processing approaches for microtubule remodeling quantification at the immunological synapse

Daniel Krentzel[a,†], Maria Isabella Gariboldi[a,†], Marie Juzans[b,c,‡], Marta Mastrogiovanni[b,c,§], Florian Mueller[a], Céline Cuche[b], Vincenzo Di Bartolo[b,*,¶], and Andrés Alcover[b,*,¶]

[a]Institut Pasteur, Université Paris Cité, CNRS-UMR3691, Unité Imagerie et Modélisation, Paris, France
[b]Institut Pasteur, Université Paris Cité, INSERM-U1224, Unité Biologie Cellulaire des Lymphocytes, Ligue Nationale Contre le Cancer, Équipe Labellisée Ligue-2018, Paris, France
[c]Sorbonne Université Collège Doctoral, Paris, France
*Corresponding authors: e-mail address: vincenzo.di-bartolo@pasteur.fr; andres.alcover@pasteur.fr

Contents

1. Introduction		41
2. Materials		44
	2.1 Equipment	44
	2.2 Disposable materials	46
	2.3 Chemicals and biological products	46
	2.4 Software packages and requirements	48
	2.5 Human subjects and blood samples	48
3. Methods		49
	3.1 Primary CD8[+] T cell isolation, transfection and differentiation	49
	3.2 Imaging microtubules at the immunological synapse by confocal microscopy	51
	3.3 Quantifying microtubule organization on flat immunological synapses	53
4. Results and discussion		60

[†] DK and MIG contributed equally to this work.

[‡] Present address: Department of Pathology and Laboratory Medicine, Children's Hospital of Philadelphia Research Institute, Perelman School of Medicine at the University of Pennsylvania, Philadelphia, PA, USA.

[§] Present address: Department of Developmental & Molecular Biology, Albert Einstein College of Medicine, The Bronx, New York, USA.

[¶] VDB and AA: co-senior authors.

Methods in Cell Biology, Volume 193
ISSN 0091-679X
https://doi.org/10.1016/bs.mcb.2024.02.036

Copyright © 2025 Elsevier Inc.
All rights are reserved, including those
for text and data mining, AI training,
and similar technologies.

5. Concluding remarks	64
6. Notes	64
Acknowledgments	65
Declaration of interests	66
References	66

Abstract

Immunological synapses result from a T cell polarization process, requiring cytoskeleton remodeling. Actin and microtubules drive synapse architecture and the localization of intracellular organelles, including Golgi and endolysosomal compartments, ensuring the directional localization of synapse components. Microtubule remodeling includes the centrosome polarization and the formation of a radial microtubules network, extending from the centrosome to the synapse periphery. Concomitantly, a ring of filamentous actin forms at the synapse periphery. Microtubule and actin remodeling facilitate vesicle fusion at the synapse, enabling T cell effector functions. Analyzing structural subtleties of cytoskeleton remodeling at the immunological synapse is crucial to understand its role in T cell functions. It may also pinpoint pathological states related with cytoskeletal dysfunctions. Quantifying filamentous protein network properties is challenging due to their complex and heterogeneous architectures and the inherent difficulty of segmenting individual filaments. Here, we describe the development of an image processing approach aimed at quantifying microtubule organization at the immunological synapse without the need for filament segmentation. The method is based on the analysis of the spatial and directional organization of microtubules growing from the centrosome to the synapse periphery. It is applied to investigate the importance of Adenomatous polyposis coli (Apc), a polarity regulator and tumor suppressor, in immunological synapse structure and functions and its potential implication in anti-tumor immune responses. We provide an open-source napari plugin of the outlined methods for analyzing filamentous networks.

Abbreviations

Apc, APC	adenomatous polyposis coli
BSA	bovine serum albumin
CTL	cytotoxic T lymphocytes; cytotoxic T cells
FAP	familial adenomatous polyposis
FBS	fetal bovine serum
NA	numerical aperture
PBMC	peripheral blood mononuclear cell
PBS	phosphate buffer saline
PFA	paraformaldehyde
siRNA	small interfering RNA
TCR	T cell receptor

1. Introduction

Immunological synapse formation relies on a complex process of T cell polarization towards antigen presenting cells or target cells. T cell polarization is initiated by T cell antigen receptor (TCR) engagement and downstream signal transduction and involves an array of signaling, cytoskeleton and vesicle traffic regulators (Mastrogiovanni, Juzans, Alcover, & Di Bartolo, 2020). Actin and microtubule cytoskeleton remodeling mediates the translocation of several organelles to the antigen-presenting cell contact site. Thus, the Golgi apparatus, as well as the endosomal and lysosomal compartments reorient their intracellular vesicle traffic towards the immunological synapse facilitating T cell effector functions.

Cytotoxic T lymphocyte (CTL) effector function relies on the appropriate delivery of lytic granules to the plasma membrane and on their fusion at the zone of contact with target cells. Lytic granules are part of the late endosomal/lysosomal compartment. They use molecular motors and intracellular traffic regulators to reach, dock and fuse with the plasma membrane. The delivery of lytic granule components (e.g., perforin and granzymes) alongside death receptor signals provokes target cell death (de Saint Basile, Menasche, & Fischer, 2010). Centrosome polarization and docking to the immunological synapse, as well as actin polymerization and clearance from the center of the contact site facilitate the formation of a secretory domain. This domain allows optimal lytic granule delivery and fusion at the CTL-target cell contact (Ritter et al., 2015; Stinchcombe, Bossi, Booth, & Griffiths, 2001; Stinchcombe, Majorovits, Bossi, Fuller, & Griffiths, 2006; Tamzalit et al., 2019). However, it is likely that target lysis can be induced by other mechanisms, since it also occurs outside the immunological synapse and independently of centrosome polarization (Bertrand et al., 2013; Tamzalit et al., 2020). Conversely, filamentous actin recovery at the center of the immunological synapse associates with the termination of lytic granule release (Ritter et al., 2017). Defects in regulators of vesicle traffic or cytoskeleton dynamics are associated with impaired CTL function and immunodeficiency (de Saint Basile et al., 2010; Dupre, Boztug, & Pfajfer, 2021; Randzavola et al., 2019).

The involvement of cell polarity regulators in this process has been recently unveiled (Juzans et al., 2020; Mastrogiovanni, Di Bartolo, & Alcover, 2022; Mastrogiovanni et al., 2022). Cell polarity regulators are scaffold proteins, displaying an array of protein-protein interaction motifs. Their effectors are involved in mechanisms controlling cytoskeleton organization, cell shape and symmetry. Among them, *Discs large homolog 1* (Dlg1) and *Adenomatous polyposis coli* (Apc) play key roles in immunological synapse formation and function in CD4 and CD8 T cells. These regulators control actin and microtubule cytoskeleton organization at the synapse and T cell activation (Aguera-Gonzalez et al., 2017; Cuche et al., 2023; Juzans et al., 2020; Lasserre et al., 2010; Round et al., 2007, 2005).

In vitro studies have been used to explore the role of cell polarity regulators in synapse formation and stability mediated by actin and microtubule cytoskeleton organization. In these studies, anti-CD3 antibody coatings of glass coverslips were used to induce the formation of pseudo-immunological synapses (Aguera-Gonzalez et al., 2017; Cuche et al., 2023; Juzans et al., 2020; Lasserre et al., 2010). Healthy cells typically form radially organized microtubules at the cell-coverslip interface indicating synapse stability, while mutations of polarity regulators have been shown to result in more disorganized phenotypes. Quantifying these differences in cytoskeletal organization is challenging due to the complexity of microtubule network morphologies and prior work in this field has relied on visually distinguishing between cells with organized and disorganized microtubule phenotypes (Cuche et al., 2023; Juzans et al., 2020). Image analysis methods present a promising avenue for more objectively quantifying these differences.

Several tools have been developed for image-based filamentous protein quantification. A key aspect of many image analysis tools is segmentation, that is the process of determining which pixels pertain to a specific category of interest. If filamentous proteins are quantified inside cells, segmenting individual cells is also necessary to assign quantified fibers to individual cells. To assess the fibers themselves, whether within cells or deposited on surfaces, two broad approaches can be employed: segmenting individual fibers to directly measure their properties, or inferring their properties indirectly from their associated signal (e.g., from fluorescent staining).

An example of a segmentation-based approach is FibrilJ, an ImageJ plugin (https://doi.org/10.17632/ndxb93h4vc.1). It measures the distribution of fibril properties, such as length, diameter and persistence length (a metric that relates to the polymer bending stiffness) from atomic force microscopy or electron microscopy images of fibrils assembled on various types of surfaces (Sokolov, Belousov, Bondarev, Zhouravleva, & Kasyanenko, 2017). FibrilJ extracts these properties by first generating binary masks of the fibers and then analyzing them using skeletonization (https://imagej.net/plugins/analyze-skeleton/).

MicroFilament Analyzer (https://www.uantwerpen.be/en/research-groups/bimef/downloads/microfilament-analyzer/) is a software package developed by Jacques et al. (2013) to extract the main orientation directions of filaments. It was validated for the study of microtubule organization in *Arabidopsis thaliana* cells. MicroFilament Analyzer detects individual cells and analyzes filament orientation with a "virtual rotating polarizer." It computes the overlap between the fibrils and a grid of lines that is rotated to detect filaments at different angular orientations.

ImageJ Directionality plugin (http://imagej.net/Directionality), on the other hand, generates histograms of structure orientations in images based on either Fourier component analysis or on the local gradient orientation calculated through a Sobel filter. OrientationJ (http://bigwww.epfl.ch/demo/orientationj) is another ImageJ software package for quantification of parameters such as local angular distribution, waviness and orientation based on structure tensors, which are matrix representations of partial spatial derivatives of the image. OrientationJ also enables visualization of orientations as well as metrics related to isotropicity, such as coherence (Rezakhaniha et al., 2012). Details of the underlying methodology and the mathematical basis of different metrics outputted by this approach can be found in the original publications by Puspoki, Storath, Sage, and Unser (2016) and Rezakhaniha et al. (2012). PyTextureAnalysis (https://github.com/ajinkya-kulkarni/PyTextureAnalysis) and OrientationPy (https://pypi.org/project/orientationpy) are Python implementations of the method outlined in Puspoki et al. (2016).

Radial structures, however, pose specific challenges as no dominant signal direction can necessarily be expected. In the context of radially organized structures (Tee et al., 2023) assessed the emergence of chirality in the actin cytoskeleton of micropatterned cells. Fluorescently labeled actin fibers were identified using a Unet-ResNet50 deep learning model and skeletonized. Cell masks consisting of concentric "rings" were generated through fixed width areas starting at the cell edge. These rings were used to calculate the tilt of fibers (that is the angle offset from radial) in different rings. The radial fiber tilt angle profile at different distances from the cell edge was then used to assess chirality.

However, segmenting individual filamentous proteins inside cells is challenging and often prone to error, given their thin structure, the presence of overlaps and their architectural complexity. Here, we present a fluorescent signal-based approach that enables extracting meaningful metrics from filamentous structures within individual cells, while circumventing the need to segment individual microtubules. Namely, we propose two metrics: the "degree of radiality" (DoR), measuring the degree to which microtubules organize radially from the centrosome, and the signal sum "skewness," capturing the mass distribution radial symmetry of microtubules, as descriptors of microtubule organization within the cell. Inspired by Tee et al. (2023), we developed a cell contour- and centrosome-aware annular subsectioning approach to assess how these metrics change across subcellular regions.

To demonstrate this approach, we investigated in a quantitative manner microtubule patterns at the immunological synapse of CTLs that were perturbed upon Apc silencing and in T cells from familial adenomatous polyposis (FAP) subjects carrying Apc mutations compared to control cells from healthy subjects. We have shown before that these synapses clearly display finely organized microtubule networks and centrosome polarization that is disturbed under Ezrin, Dlg1 and APC protein deficiencies. We had observed differences through visual appreciation of anonymized cell images (Aguera-Gonzalez et al., 2017; Juzans et al., 2020; Lasserre et al., 2010). Our novel quantitative approach was applied to the same images comprising flat and spread two-dimensional pseudo-immunological synapses on which accurate confocal imaging could be performed and microtubule patterns were well defined (Fig. 1). To benchmark this more objective approach, we compare quantifications to manual cell annotations by trained immunologists.

2. Materials

2.1 Equipment

1. Cell culture incubator allowing standard cell culture conditions in a humidified atmosphere (37 °C, 5% CO_2).
2. Standard bench top centrifuge (Eppendorf Centrifuge 5810R or 5415R).
3. Sterile cell culture laminar flow hood with safety level II.
4. MACS MultiStand™ (Miltenyi Biotec, No 130-042-303).

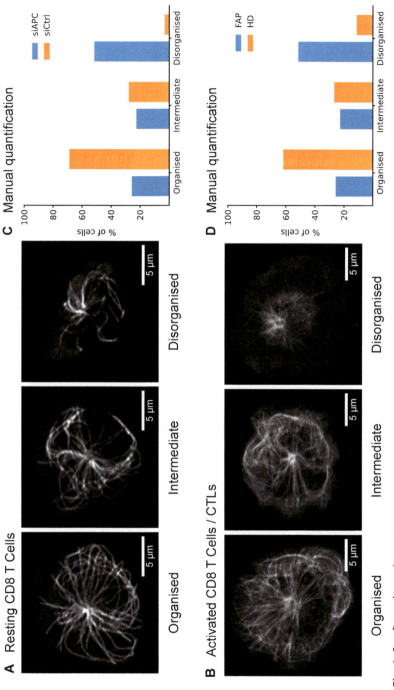

Fig. 1 See figure legend on next page.

5. MidiMACS™ Separator (Miltenyi Biotec, No 130-042-302).
6. Biosafety class 2 laboratory (BSL2 or P2).
7. Microscope: LSM 700 confocal microscope (Zeiss) using the Plan-Apochromat 63×/1.40 Numerical aperture (NA) objective and ZEN software (Zeiss).

2.2 Disposable materials

1. 24-Well plates for cell culture (Falcon, No 353047).
2. Sterile microcentrifuge tubes (Eppendorf No 3810).
3. 96-Well plates for cell culture (TPP, No 92097).
4. Coverslips

2.3 Chemicals and biological products

1. Lymphocyte Separation Medium Pancoll Human tubes (Pan Biotech, No P04-60125).
2. Magnetic cell sorting CD8+ T cell isolation kit (Miltenyi Biotec, No 130-096-495).
3. LS columns for magnetic cell sorting (Miltenyi Biotec, No 130-042-401).
4. RPMI 1640 cell culture medium containing GlutaMAX-I and Phenol Red (Gibco, Thermo Fisher Scientific, No 61870).
5. Sodium pyruvate (Life Technologies, No 11360).
6. Nonessential amino acids (Life Technologies, No 11140).

Fig. 1 Differential microtubule network organization at the immunological synapse. Involvement of the polarity regulator APC. Examples of organized (left panels), disorganized (right panels) or intermediate (middle panels) microtubule patterns observed at the immunological synapse formed by T cells spread on anti-CD3-coated coverslips. A projection of the 4 confocal optical sections closest to the coverslip are shown. (A) Resting human CD8+ T cells transfected with siRNA control (siCTRL) or Apc (siAPC). (B) Activated CD8+ T cells from FAP and healthy subjects differentiated *in vitro* into CTLs. Please note that images displayed are examples that may be found in both siCtrl and siRNA cells or in healthy and FAP subject cells. (C, D) Visual ranking into organized, disorganized or intermediate phenotypes performed by one (C) or three (D) independent investigators and assigned based on majority vote. A similar analysis on the experiments with siRNA-transfected cells, carried out by three independent investigators using anonymized images, has been previously reported (Juzans et al., 2020). Data from FAP and healthy patients used here (B, D) are a subset of those described in Cuche et al. (2023).

7. HEPES (*N*-(2-hydroxyethyl)piperazine-*N'*-(2-ethanesulfonic acid) (Life Technologies, No 15630-056).
8. Penicillin–streptomycin (Gibco, No 15140-122).
9. Human serum (Dominique Dutscher, No S4190-100).
10. Fetal bovine serum (FBS): HyClone™ Serum—Research grade fetal bovine serum, origin South America (Dominique Dutscher, No SV30160.03).
11. Culture medium for T cells: RPMI-1640 + GlutaMAX-I and PhenolRed supplemented with 5% human serum, 1 mM Sodium Pyruvate, 1× non-essential amino acids, 10 mM HEPES and 5 U/mL penicillin–streptomycin.
12. Anti-human CD3ε antibody, clone UCHT1 (BioLegend Inc., No 300402).
13. Anti-human CD28 antibody (Beckman Coulter, No IM1376).
14. Recombinant human IL-2 (PeproTech, No 200-02).
15. Dulbecco's PBS-Modified, w/o $CaCl_2$ and $MgCl_2$ (DPBS) (Gibco, Thermo Fisher Scientific, No 14190).
16. Puromycin (Gibco, No A11138-03).
17. Bovine serum albumin (BSA) (Alpha Diagnostics, No 80400-100).
18. Fixable Viability Stain 450 (BD Biosciences).
19. Paraformaldehyde (PFA) solution (4% in PBS) (EMS, No 15714-S).
20. Poly-L-lysine solution 0.1% (w/v) (Sigma-Aldrich, No P8920).
21. Saponin (Sigma, No S7900).
22. Triton X-100 (Sigma, No T9284).
23. HCl 1 N (Fluka, No. 84436) solution in 70% ethanol (Fisher Chemical No E/0550DF/21).
24. Antibodies for FACS detection of CTL: anti-CD3-PE-Cy5 (1/30; clone HIT3a; BD Biosciences), anti-CD8-APC-Cy7 (1/30; clone RPA-T8; Biolegend).
25. Anti-Granzyme-B-PE antibody (1/50; clone GB12; Invitrogen).
26. Antibodies for microscopy: anti-β-tubulin (6.6 μg/mL; Merck Millipore) and anti-pericentrin (1/100; Abcam).
27. Small interfering RNA (siRNAs) oligonucleotides targeting Apc and negative control were previously described (Juzans et al., 2020). Double stranded RNA oligonucleotides sequences were used. For Apc depletion siRNA sequence was 5′-GAGAAUACGUCCACACCUU-3′ (siApc; GE Healthcare). Control oligonucleotide sequence was 5′-UAGCGACUAAACACAUCAA-3′ (siGENOME Non-Targeting siRNA #1; GE Healthcare).

2.4 Software packages and requirements

1. FlowJo v10 software (FlowJo, LLC) for flow cytometry data analysis.
2. ImageJ (FIJI) software.
3. Huygens Pro Software (Scientific Volume Imaging) for image deconvolution.
4. MACSQuant® Analyzer flow cytometer (Miltenyi Biotec).
5. Python version 3.10.8 and packages listed in Table 1.

2.5 Human subjects and blood samples

Human peripheral blood T cells from healthy volunteers intended for Apc silencing experiments were obtained from the French Blood Bank Organization (Etablissement Français du Sang, EFS), under Institut Pasteur-EFS institutional convention.

For experiments comparing T cells from FAP subjects to healthy control subjects, we recruited FAP subjects with diagnosed mutations in the *APC* gene and the same number of sex- and age-matched (within a 0–5-year difference) healthy subjects through the Clinical Investigation and Access to Biological Resources (ICAReB-Clin) Institut Pasteur core facility (NSF 96-900 certified, from sampling to distribution, reference BB-0033-00062/

Table 1 Software packages.

Package name	Citation	URL
OpenCV	Bradski (2000)	https://github.com/opencv/opencv-python
NumPy	Harris et al. (2020)	https://github.com/numpy/numpy
Matplotlib	Hunter (2007)	https://github.com/matplotlib/matplotlib
Scikit-image	van der Walt, Nunez-Iglesias, Boulogne, Warner, and Yager (2014)	https://github.com/scikit-image/scikit-image
Tifffile		https://github.com/cgohlke/tifffile
SciPy	Virtanen et al. (2020)	https://github.com/scipy/scipy
Pyefd	Kuhl and Giardina (1982)	https://github.com/hbldh/pyefd
Napari	Sofroniew (2009)	https://github.com/Napari/napari

ICAReB platform/Institut Pasteur, Paris, France/BBMRI AO203/1 distribution/access: 2016, May 19th, [BIORESOURCE]). Recruitment and clinical research protocols were approved by a national ethical committee (*Comité de Protection des Personnes, Île de France-1*), CoSImmGEn cohorts Protocol N° 2010-déc. 12483 for healthy subjects and N° 2018-mai-14852 for FAP subjects. FAP subjects were recruited through the *Association Polyposes Familiales France*. ICAReB-Clin ensured all subject visits. All FAP and healthy subjects signed informed consent. Non-inclusion criteria for all subjects were to have a pathology or a treatment potentially affecting immune responses at the time or within the 2 weeks preceding the visit. Each subject filled together with the clinical research physician at ICAReB-Clin a health questionnaire. For FAP subjects, this questionnaire included the recording of the mutation type documented by the genetic analysis report extracted from their medical file. A clinical blood analysis was performed at each visit. Subjects were 7 women and 5 men. Average age was 50.4 years for healthy subjects (range 30–69) and 50.1 years for FAP subjects (range 29–72) with an average age matching difference of 2.6 years. APC mutations, according to the genetic diagnosis report, included frameshift mutations potentially leading to lack of protein expression of the mutated allele or expression of truncated forms.

Blood samples, 100 mL per subject, from both FAP and healthy subjects were withdrawn from the antecubital vein and collected in sodium-heparin tubes.

3. Methods

3.1 Primary CD8$^+$ T cell isolation, transfection and differentiation

3.1.1 Peripheral blood mononuclear cell (PBMC) isolation

Separate PBMC by density gradient centrifugation using Lymphocyte Separation Medium Pancoll Human (see **Note 1**).

3.1.2 CD8$^+$ T cell isolation by magnetic cell sorting

1. Isolate CD8$^+$ T cells from PBMC by negative selection using magnetic cell sorting using a CD8$^+$ T Cell Isolation Kit and LS columns, following manufacturer's instructions.

2. Culture purified cells in RPMI 1640 cell culture medium supplemented with 10% FBS, 1 mM sodium pyruvate, 1% (v/v) nonessential amino acids, 10 mM HEPES, 0.5% (v/v) penicillin–streptomycin.

3.1.3 Apc silencing in resting T cells by siRNA oligonucleotides transfection

1. Freshly isolated CD8 T cells (10^7) were transfected with 1 nmol of siCtrl or siApc using the Human T Cell Nucleofector kit and the program U-14 on an Amaxa Nucleofector II (Lonza).
2. Cells were then harvested in human CD8 medium without penicillin–streptomycin and used 72 h after transfection.

3.1.4 Activation of FAP or control CD8$^+$ T cells and differentiation into CTLs

1. Coat 24-well plates with 400 µL mouse anti-CD3 (human) (10 µg/mL) in PBS overnight at 4 °C or 3 h at 37 °C.
2. Wash plates 3 times with PBS, then incubate 1 h at 37 °C with 500 µL culture medium to prevent nonspecific cell binding.
3. Resuspend freshly isolated CD8$^+$ T cells at 2×10^6 cell/mL in culture medium (see **Note 2**) containing 7 µg/mL anti-CD28 (note that anti-CD28 is used soluble, whereas anti-CD3 is coated to the plate) and 100 U/mL recombinant human IL-2. Remove medium from plates and distribute 500 µL of cell suspension in each well, incubate plates at 37 °C with 5% CO_2 for 2 days (infection) or 6 days (FACS analysis). For FACS analysis, the concentration is adjusted to 2×10^6 cell/mL in culture medium containing 100 U/mL recombinant human IL-2 at days 2 and 4.

3.1.5 Assess CTL differentiation by FACS analysis

1. Take 0.5×10^6 CD8$^+$ T cells at days 0, 2, 4 and 6 and place them in a 96 well plate.
2. Wash cells twice with PBS, centrifuge plates for 6 min at $450 \times g$ remove supernatant and add 100 µL of PBS $1\times$ with Fixable Viability Stain 450 (250 ng/mL; BD) diluted 1/1000 (v/v), incubate 10 min at RT.
3. Wash cells twice with PBS, 0.5% (w/v) BSA, spin plates for 6 min at $450 \times g$, remove supernatant and add 75 µL of PBS, 0.5% BSA with staining antibodies, incubate 30 min at 4 °C.
4. Wash cells twice with PBS, 0.5% BSA, spin plates for 6 min at $450 \times g$, remove supernatant.
5. Wash cells twice with PBS, 0.5% BSA, spin plates for 6 min at $450 \times g$, remove supernatant, fix with 200 µL of PBS, 0.5% BSA, 1% PFA for 30 min at RT.

6. Wash cells twice with PBS, 0.5% BSA, spin plates for 6 min at $450 \times g$, remove supernatant and add 60 μL of PBS, 0.5% BSA, 0.05% saponin to permeabilize cell membrane, incubate 10 min at RT.
7. Without washing add 15 μL of anti-Granzyme-B-PE (1/10 v/v; clone GB12; Invitrogen), incubate 30 min at 4 °C.
8. Analyze samples by flow cytometry using a MACSQuant® Analyzer and FlowJo v10 software. All samples were gated on forward and side scatter (FSC/SSC), for single cells, and live cells.

3.2 Imaging microtubules at the immunological synapse by confocal microscopy

Observation of microtubule patterns is easier when imaging flat immunological pseudo synapses formed by T cells stimulated on anti-CD3-coated coverslips.

3.2.1 Coat glass coverslips with anti-CD3 antibody

1. Wash glass coverslips with HCl-EtOH 70% for 10 min, rinse twice with water and once with EtOH 70% before letting them dry.
2. Coat glass coverslips with poly-L-lysine at 0.002% (w/v) in water for 30 min at room temperature, in 24 well plates, wash once with water, let them dry (see **Note 3**).
3. Coat poly-L-lysine-coated coverslips with anti-CD3 antibody at 10 μg/mL in PBS 3 h at 37 °C or overnight at 4 °C, wash 3 times with PBS, incubate 1 h at 37 °C with culture medium to prevent cell non-specific binding (see **Note 3**).

3.2.2 Immunological synapse formation, fixation and immunofluorescence staining

1. Plate T cells on coverslips for 5 min at 37 °C.
2. Fix with 4% paraformaldehyde for 13 min at room temperature. For microtubule detection, incubate cells 20 min at -20 °C in ice cold methanol.
3. Wash coverslips in PBS and incubate them 1 h in PBS with 1% bovine serum albumin (BSA) (v/v) to prevent nonspecific binding.
4. Incubate 1 h at room temperature with PBS, 1% BSA, 0.1% Triton X-100 and anti-β-tubulin (6.6 μg/mL; Merck Millipore) and anti-pericentrin (1/100; Abcam) antibodies (see **Note 4**). Pericentrin is localized at the centrosome, and it is a good marker for centrosome and microtubule organizing center.

5. Incubate coverslips with the corresponding fluorescent-coupled secondary Ab for 45 min at room temperature and then wash three times in PBS with 1% BSA (see **Note 4**).

6. Mount coverslips on microscope slides using ProLong Gold Antifade mounting medium with DAPI (Life Technologies) to stain nuclei, leave dry overnight and seal with nail polish to prevent drying.

7. Store coverslips at 4 °C until microscope observation (see **Note 5**).

3.2.3 Confocal microscopy and image pre-processing

1. Confocal images were acquired with an LSM 700 confocal microscope (Zeiss) using the Plan-Apochromat 63×/1.40 NA objective. Optical confocal sections were acquired at 0.2 μm intervals using ZEN software (Zeiss) by intercalating green and red laser excitation to minimize channel cross talk.

2. For manual classification of cell phenotypes, images were treated by deconvolution using Huygens Professional Software (Scientific Volume Imaging, v14.10). Source images had a pixel size of $66 \times 66 \times 200$ nm and had no intensity saturation. In Huygens, we used the Classical Maximum Likelihood Estimation deconvolution algorithm, configured with the following parameters: maximum number of iterations: 50; threshold on quality variation: 5%; iteration mode: optimized; and brick layout processing algorithm: automatic. The deconvolution process converged typically in 30–50 iterations. Images used for the automated analysis pipeline were not deconvolved.

3.2.4 Manual classification of cell phenotype

1. Morphological phenotypes of human resting CD8 T cells transfected with siCtrl or siAPC oligonucleotides had previously been classified by observation of deconvolved maximum intensity projections of four confocal sections close to the coverslip surface by three different investigators (Juzans et al., 2020). This classification was repeated by one investigator on deconvolved maximum intensity projections based on the same phenotype definition but with the addition of a third class (intermediate appearance between organized and disorganized phenotypes) to account for more nuanced phenotypes. Images were analyzed using ImageJ (Fiji) software. Examples of these three phenotypes and resulting quantification results can be seen in Fig. 1A and C respectively.

2. Images from healthy and FAP subject experiments were similarly analyzed based on the same three phenotype definitions. For this experimental setup, 4 confocal sections closest to the surface were projected

after Huygens deconvolution to exclude microtubules further from the coverslip interface. Examples of the three phenotypes and relative manual classification results can be seen in Fig. 1B and D respectively.

3.3 Quantifying microtubule organization on flat immunological synapses

3.3.1 Quantitative analysis pipeline of microtubule radiality and mass distribution asymmetry

To quantify differential microtubule patterns, we developed an image analysis toolkit implemented in Python version 3.10.8. This approach assumes that two important features for quantifying the level of organization of the microtubule network are (1) the degree to which microtubules are radial from the centrosome as an organizing center or from the geometric center of the convex hull of the cell, captured by the "degree of radiality" (DoR) and (2) the degree to which microtubules are distributed in a radially symmetric manner within the cell, captured by the "skewness" of the signal sum projection. This toolkit can be used to quantify microtubule properties of T cells spread on anti-CD3-coated coverslips but may also be a useful starting point to assess filamentous protein organization inside other cell types. The method can be replicated through a napari plugin (https://github.com/krentzd/napari-microtubule-analyzer) and comprises the steps below, summarized graphically in Fig. 2. The interface of the plugin and relevant functionalities and modifiable parameters are shown in Fig. 3. Raw (non-deconvolved) image stacks were used for all analyses described below.

1. For DoR quantification, maximum intensity projections of the β-tubulin signal were generated over the whole stack for images obtained in silencing experiments. For FAP subject experiments, only confocal sections closest to the coverslips were used to exclude microtubules far from the pseudo-synapse. To automate the projection of slices closest to the coverslip in a reproducible and objective manner, the slice with the highest sum intensity was identified. The maximum projection was then obtained from the range of slices just below and above this slice (three total slices). For the skewness quantification, sum projections of the β-tubulin signal were obtained across the whole stack of images for both experimental setups. Where centrosome staining was available, sum intensity projections of the pericentrin signal were generated to obtain the centrosome position.

2. Segmentation masks of individual cells were obtained with Otsu's method (Otsu, 1979) applied to the maximum intensity projection or sum projection (as described in step 1) of the β-tubulin signal followed

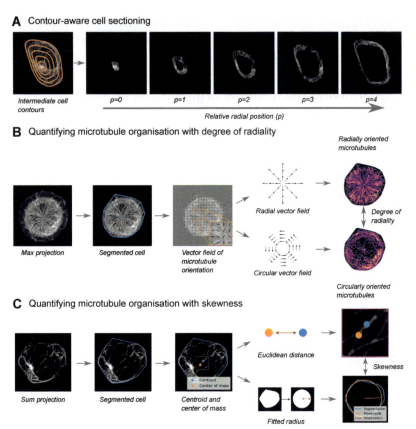

Fig. 2 Microtubule network analysis pipeline approach. (A) Cell contour- and centrosome-aware cellular sectioning based on elliptical Fourier descriptors (EFDs). This approach enables generating intermediate contours starting from a small circle enclosing the centrosome increasingly becoming more similar to the contour of the cell convex hull as they approach the cell perimeter. (B) Graphical representation of the approach used for microtubule degree of radiality (DoR) quantification. Vector fields are computed on maximum intensity projections of microtubule networks and radial and circular vector fields are generated centered at the centrosome or geometric center of the convex hull. The DoR is then calculated as the sum of the dot product of the microtubule vector field and the radial vector field divided by the sum of the dot product of the microtubule signal vector field and the circular vector field. Centrosome and convex hull outlines are shown in blue. (C) Graphical representation of microtubule skewness quantification. The center of mass (orange) and centroid (blue) are computed for the sum projection of the segmented cell. The Euclidean distance between the centroid and the center of mass is calculated and normalized to the radius of a circle with the same area as the cell convex hull. Approaches are illustrated for entire cells but can be applied to each cellular subsection.

Quantifying microtubule remodeling at the immunological synapse by image analysis 55

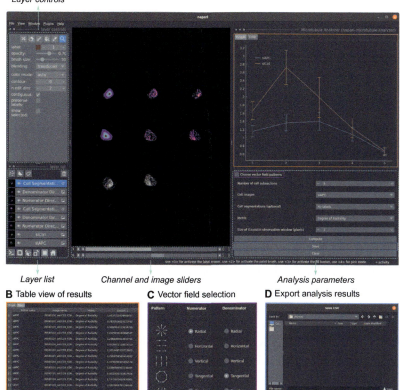

Fig. 3 Components of the napari plugin interface. Once installed, the plugin can be accessed by selecting "Plugins > Microtubule Analyzer. (napari-microtubule-analyzer)." (A) Images are represented as "layers" in napari which can be displayed or hidden by toggling the eye icon in the "layer list." Layers can be re-arranged by dragging and dropping and properties of selected layers can be modified through the "layer properties" panel. The displayed images and channels are chosen with image and channel sliders. The napari plugin is displayed on the right-hand-side of the interface and analysis parameters—like the metric (DoR or skewness), number of subsections or the size of the Gaussian observation window—can be modified. To analyze a dataset, the folder containing the images can be dragged and dropped directly into the viewer. The desired dataset can then be chosen from a drop-down menu ("Cell images") and if available, cell segmentation masks can be provided ("Cell segmentations"). Segmentation masks and filtered images are displayed in the napari viewer. Results are then displayed graphically and in tabular form (B). The table enables navigating to specific cells by clicking on the relevant row in the table. (C) The vector field pattern can also be modified to account for different phenotypes. (D) Analysis results can then be saved as a CSV file for downstream analysis. Sample siRNA data can be loaded by selecting "File > open sample—Sample siRNA data (napari-microtubule-analyzer)." Detailed documentation can be found on the GitHub plugin repository (https://github.com/krentzd/napari-microtubule-analyzer).

by a binary opening step. Cells touching the image border were removed. The centrosome was identified by computing the median x and y positions (in case multiple pixels had the same maximum value) of the peak signal in the sum intensity projection of the pericentrin signal. Where centrosome staining was not available, the geometric center of the convex hull was utilized as an approximation. When both these positions were available, their Euclidean distance normalized to the fitted radius (radius of a circle with the same area as the convex hull) was 0.250 ± 0.148 for siCtrl cells and 0.256 ± 0.164 for siAPC cells (mean $\pm$ standard deviation). This data indicates that the distance between the two points is comparable in both conditions and that the geometric center of the convex hull is a reasonable alternative to the centrosome for the proposed methodological approach and samples.

3. A convex hull was generated for binarized microtubule structures to estimate the area taken up by the cell. This approach assumes that the cell perimeter is convex and that the microtubules spread up to the periphery of the cell. Alternatively, the cell surface area can be extracted from fluorescently stained membrane proteins such as CD3, when available.

4. Elliptical Fourier Descriptors (EFDs) (Kuhl & Giardina, 1982) were used to generate cellular subsections. Briefly, EFDs allow to represent a contour as a sum of ellipses, with a higher number of harmonics capturing more fine-grained details of the structure. More details about how EFDs are computed by the pyefd package can be found in the documentation (https://pyefd.readthedocs.io/en/latest/). Contour-aware cellular subsections were obtained as follows. EFDs were used to fit the contour of the convex hull using 10 harmonics (hereafter referred to as the outer contour). A small circular contour was fitted to the centroid (radius = 1) using the skimage.morphology.disk function (hereafter referred to as the inner contour). Different numbers of intermediate outlines based on the number of expected sections can be obtained by computing the weighted average of the inner and outer contour EFD coefficients. The spaces between these contours can then be used as masks to evaluate downstream metrics in different subcellular annular regions.

5. Cells for which segmentation failed (e.g., portions of the cells were not included in the convex hull) or automated projection failed (e.g., peak intensity profiles did not correspond to the coverslip interface) based on visual inspection were excluded from the analysis.

6. The **degree of radiality (DoR)** captures the ratio between the radial and the circular signal, with values higher than 1.0 signifying

predominantly radial organization and values lower than 1.0 indicating predominantly circular organization of fluorescent structures. Its calculation was performed as follows:

a. Radial and tangential (hereafter referred to as "circular") unit vector fields were generated at each pixel using the centrosome as an origin. Where pericentrin staining was not available, the geometric center of the convex hull was used as an origin instead (see **Note 6**). In different model cell types, different vector fields might be appropriate based on the filamentous protein organization (e.g., horizontal), with the caveat that the two vector fields should be perpendicular to each other.

b. Microtubule directionality vectors were extracted from the Gaussian filtered ($\sigma = 0.1$) maximum intensity projection by estimating the dominant local orientation using a structure tensor (Puspoki et al., 2016). First, local image gradients were obtained in the x- and y-direction using a Sobel filter from which a structure tensor was constructed using a Gaussian observation window size of 1 pixel. Then, a vector field was obtained from the computed structure tensor for pixels above the half median intensity value of the image to reduce the influence of background signal. The code for computing the structure tensor and obtaining the vector field is based on the source code of PyTextureAnalysis.

c. The absolute dot product was then computed between the vector fields obtained from the structure tensors and the radial and circular vector fields respectively (see **Note 7**).

d. The degree of radiality (DoR) is defined as the ratio of the sum of the absolute dot products of the structure tensor and the radial vector field, and the sum of the absolute dot product of the structure tensor field and the circular vector field. The DoR was evaluated for each cellular subsection independently by making use of the masks obtained in step 4.

7. The **microtubule skewness** measures the signal distribution's radial symmetry, with a value of zero indicating a radially symmetric signal and higher values (with a theoretical maximum value around 1.0) denoting radially unbalanced signal distribution. This analysis was performed as follows:

a. To account for all the β-tubulin signal within the cell, the sum, rather than the maximum, intensity projection of the β-tubulin signal was obtained and used to identify the "center of mass" of the cell.

b. The distance between the center of mass position, which is affected by the spatial distribution of the signal, and the centroid, which represents the geometric center of the cell, was used to assess how skewed the microtubule network is within the cell. This distance was normalized to the fitted radius of the cell (the radius of a circle with the same area as the convex hull) to account for differences in overall cell size. This analysis was performed for each cellular subsection individually.

Detailed installation and usage instructions for use of the napari plugin can be found at https://github.com/krentzd/napari-microtubule-analyzer along with a set of sample images. An explanation of the interface can be found in Fig. 3. Briefly, the plugin allows users to load data (and custom segmentations if needed) and run either the DoR or skewness analysis by selecting the preferred vector field patterns (radial, horizontal, vertical, tangential or random) for the numerator and denominator, the number of subcellular slices and the Gaussian observation window size. The plugin assumes input images are maximum intensity or sum projections and utilizes the geometric center of the cell segmentation for cellular subsectioning unless a second channel containing centrosome staining is provided. The data output, subcellular segmentations, the filtered signal images (signal dot product with each vector field direction) and graphs can then be exported. To tune a broader set of parameters, users are encouraged to refer to the plugin source code.

3.3.2 Application of the analysis pipeline to lymphocytes from Apc silencing and FAP subject experiments

To validate the analysis pipeline, the above steps were applied to data from silencing and FAP subject experiments and benchmarked against annotations of cells by visual inspection. All images used were non-deconvolved 32-bit images and cells were evaluated using 5 subsections. This number was chosen as it was empirically found to discriminate well between cell populations of interest and recover trends observed qualitatively. For DoR analyses (Fig. 4A and B), maximum intensity projections of whole stacks were used for silencing experiments, whereas only three z-slices at the interface between the cell and the coverslip were used for maximum intensity projections for CTLs from FAP and healthy subject experiments, as described in step 1 in Section 3.3.1. This is because differentiated T cells in this experimental setup formed more complex microtubular structures beyond the synapse interface, which could obscure the relevant microtubule network.

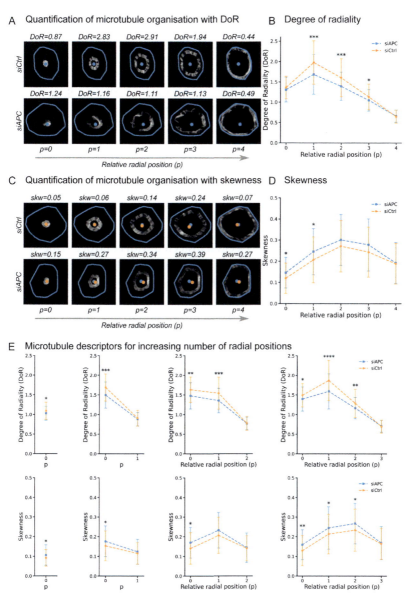

Fig. 4 Microtubule degree of radiality and mass skewness quantification in siAPC and siCtrl cells. (A, C) Visual representation of respectively the DoR and skewness computation for cellular sections at different relative positions starting from the centrosome ($P=0$). (B, D) DoR and skewness distribution profiles for each cellular section starting at the centrosome. Similar analyses for FAP subject experiments were reported in Cuche et al. (2023). (E) Variation in DoR (top) and skewness (bottom) profiles for different numbers of cellular sections. siCtrl: $n=101$ siAPC: $n=92$. In (A), blue dots correspond to the centrosome. In (C), blue dots correspond to the centroid and orange dots to the center of mass. Details on the data pre-processing can be found in Section 3.3.2. Significance was determined by Student's t-test. *$P \leq 0.05$, **$P \leq 0.01$, ***$P \leq 0.001$, ****$P \leq 0.0001$.

The slice with the highest total intensity was used as an indicator of the cell-coverslip interface (as on average, this is the slice with the cell's highest cross sectional area) and the maximum intensity projection was conducted for that slice along with the slice immediately above and the slice immediately below. For skewness analysis (Fig. 4C and D), sum projections of the entire cell stacks were used for both silencing and FAP subject experiments. Due to the absence of pericentrin staining for FAP subject experiments, the geometric center of the convex hull was used instead of the centrosome position as the "organizing center" for this dataset.

4. Results and discussion

Resting human CD8+ T cells transfected with control or Apc-specific siRNA were set to spread on anti-CD3-coated coverslips to form flat immunological pseudo-synapses that display different microtubule patterns: (i) a regular microtubule radial pattern (organized) with polarized centrosome (microtubule organizing center) close to the contact site and in the middle of a rather symmetrical cell structure, typical of control cells (Fig. 1A, left panels); (ii) irregular (disorganized) patterns displaying many more turning microtubules, reaching less often the edges of the synapse, and with less polarized centrosomes, more prevalent in APC silenced cells (Fig. 1A, right panels). Activated CD8+ T cells from healthy and FAP subjects undergoing the same stimulatory conditions exhibited different patterns but with similar characteristics, i.e., with organized patterns more frequently observed in healthy subjects versus disorganized patterns prevailing in FAP subjects (Fig. 1B, left and right panels, respectively). To account for ambiguous phenotypes, an intermediate phenotype was added for the classification (Fig. 1A and B, middle panels). Based on these reference phenotypes, cells were assessed by visual inspection by experienced immunologists to quantify the prevalence of different phenotypes in Apc silencing or FAP subject experiments (Fig. 1C and D). The data from Apc silencing experiments in resting CD8 T cells is reported in more detail in Juzans et al. (2020), where we concluded that Apc controls these patterns, since APC-silenced cells displayed more frequently disorganized patterns with less polarized centrosomes than controls. Interestingly, microtubules in resting, freshly isolated human T cells were less numerous and thicker, whereas within *in vitro*-activated, differentiated T cells, as used in FAP and healthy subject experiments, microtubules were thinner and more numerous (compare Fig. 1A and B). While visual inspection provided a useful indication of the underlying mechanistic role of Apc, this type of analysis is based on qualitative observations and therefore subjective. Further, the analysis was particularly difficult in the case of differentiated T cells from FAP and healthy subjects given their higher architectural complexity.

Therefore, we developed a method aiming to characterize distinct microtubule patterns at the immunological synapse for different experimental systems (Fig. 2). To account for intracellular heterogeneity of microtubule patterns, a contour-aware subsectioning approach relying on EFDs was developed (Fig. 2A). To assess microtubule organization, two metrics were developed: the degree to which the direction of microtubules is radial from the centrosome, captured by the degree of radiality (DoR) (Fig. 2B), and the degree to which the spatial distribution of microtubules is distributed in a radially symmetric manner in the cell, captured by the sum signal skewness (Fig. 2C).

The cellular subsectioning approach is helpful in assessing proteins involved in the cytoskeleton and structural stability of the cell, and particularly microtubules, for which the centrosome acts as an organizing center. Furthermore, in the case of the immunological synapse specifically, polarization of the centrosome towards the center and the development of radially organized microtubules from the centrosome is key to synapse stability. A napari plugin was developed to enable the application of the methodology in other contexts, with the ability to tune the vector fields to suit different expected organization patterns relevant to different biological models (Fig. 3). Appreciable differences between conditions could be observed in Apc silencing experiments in specific cellular subsections (Fig. 4A–D). These differences in both DoR and skewness were more noticeable in intermediate cellular subsections (Fig. 4B and D). This could be explained by the fact that microtubules at the centrosome (position 0) are necessarily more constrained in their organization and microtubules at the periphery of all cells tend to bend along the contour of the cell or be less distinguishable, resulting in a more circular organization, or at least a more evenly distributed signal along the cell perimeter. In fact, for both conditions, the DoR is below 1.0 for the last section, indicating the organization is more circular than radial at the cell periphery. In line with this potential explanation, the differences between populations are only appreciable when increasing the number of cellular sections at which the metric is evaluated to better isolate differences in intermediate annular sections, highlighting the importance of considering local subcellular changes when assessing these phenotypic differences (Fig. 4E).

To validate the consistency of the outputs of this approach with classification by visual inspection, the DoR and skewness profiles of the cells were assessed by disaggregating the data based on the phenotypes assigned manually by trained immunologists. In both cell populations from silencing experiments and FAP subject experiments, the DoR profiles for cells were found to capture the differences observed through qualitative inspection, with cells classified as "disorganized" having on average lower DoR scores than "organized" cells, and "intermediate" cells scoring accordingly in intermediate annular sections (Fig. 5A and B). The same trend was not clearly observable when skewness scores were disaggregated by manual labels. This suggests that both the differences between groups and the visual discrimination of microtubule network organization are likely more related to the directionality of the microtubules as opposed to the radially symmetric distribution of microtubule signal within the cell projections (Fig. 5C and D). Based on this benchmarking, the automated DoR quantification appears to best represent the observed differences between cells captured by manual annotations. By providing information from different cellular sections and enabling assessment of both directionality and spatial distribution of microtubules, this method may potentially provide more mechanistic insights compared to subjective annotations.

While this method appears to reliably quantify microtubule network characteristics, it has some limitations. Indeed, it relies on several underlying assumptions, such as the cell perimeter being approximated by the convex hull and the centrality of the centrosome as an organizing structure. The DoR quantification also excludes low intensity signal (as defined relative to the median image signal) as noise, which might lead to the exclusion of some microtubule signal from the analysis. The cell sectioning approach provides a flexible method that is adaptable to different cell shapes and centrosome positions and might be useful for other image analysis applications but assumes a gradual transition from an inner circle to the outer cell contour will represent areas of interest. Finally, this method relies on 2D projections of 3D data, and therefore may not capture any architectural contributions along the z-axis.

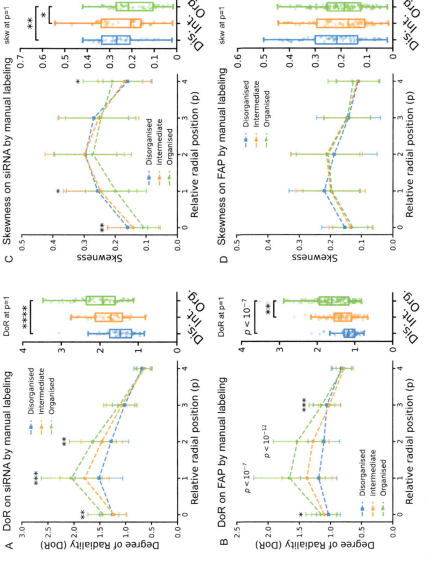

Fig. 5 See figure legend on next page.

5. Concluding remarks

Analysis of microtubule patterns is challenging due to their complexity and variability along different cellular states. The methodology described here allowed us to quantify microtubule patterns at the immunological synapse and their control by a polarity regulator. We apply the method to investigate the role of Apc, analyzing human primary T cells and their Apc-silenced counterparts, and T cells from FAP versus healthy subjects. By comparing the skewness and DoR quantification results to manual annotations, we show that the differences captured through manual annotations of cells and consequently the effect of Apc were more related to microtubule radial organization changes in the central annular subsections of the cell as opposed to changes in radial symmetry of the microtubule "mass distribution." The method can distinguish microtubule patterns previously appreciated only by visual inspection of samples (Juzans et al., 2020). It reduces the potential ambiguity and bias of visual ranking and provides similar results. However, it may be more reductionist than visual inspection in terms of analyzing more complex patterns and could thus be improved upon in future work.

6. Notes

1. Bring tubes containing the Human Lymphocyte Separation Medium Pancoll to room temperature. Carefully pour 30 mL of blood into each tube. Centrifuge at $800 \times g$ for 30 min in a centrifuge with a swing-out rotor and brake switched off. Harvest the PBMC fraction between the

Fig. 5 Manual benchmarking of the quantification approach. (A, B) DoR quantifications disaggregated by manual labeling for both silencing and FAP subject experiments. (C, D) Skewness quantifications disaggregated by manual labeling for both silencing and FAP subject experiments. Note that the cell sectioning was done from the centroid of the segmentation instead of the centrosome position for FAP data. siRNA disorganized ($n = 30$); siRNA intermediate ($n = 33$); siRNA organized ($n = 57$). FAP disorganized ($n = 53$); FAP intermediate ($n = 57$); FAP organized ($n = 128$). Details on the data pre-processing can be found in Section 3.3.2. Significance between the manually classified images was assessed with a one-way ANOVA. Following statistically significant P-values ($P \leq 0.05$), a post-hoc Tukey's HSD test was performed to assess the pairwise difference between groups.* $P \leq 0.05$, **$P \leq 0.01$, ***$P \leq 0.001$, ****$P \leq 0.0001$.

plasma and the Pancoll with a pipette as depicted in the manufacturer's instructions. The membrane present in the tubes prevents contamination with granulocytes and erythrocytes. Wash twice the lymphocytes/PBMCs with DPBS, spin 10 min at $450 \times g$.

2. The use of human serum instead of fetal calf serum strongly improves cell culture conditions and viability of human T cells.

3. Coverslips can be washed and coated with poly-L-lysine up to a week in advance in 24 well plates. However, it is better to coat them with anti-CD3 antibodies the day of the experiment.

4. To save reagents, coverslips containing cells may be stained by setting them upside down on drops of staining solution laid on parafilm.

5. For long storage, keep mounted coverslips in boxes at $-20\,^{\circ}\mathrm{C}$.

6. Cellular subsection mask generation is best done using a circle around the centrosome position as an inner contour, given the centrosome acts as the microtubule organizing center. However, this can also be done using the geometric center of the convex hull if centrosome staining is not available. In this case, radial and tangential vector fields should be generated using the geometric center of the convex hull for consistency.

7. Note that image structure tensors can only take values between -90° and $+90^{\circ}$, whereas vectors in the radial and circular vector fields are defined between 0° and 360°. The fact that the angle ranges differ, however, is not an issue since $A \cdot B = |A||B| \cos(\theta)$ and the absolute value of the cosine of an angle is the same as the absolute value of its supplementary angle.

Acknowledgments

This work was supported by a grant from La Ligue Contre le Cancer Equipe Labellisée Ligue-2018 and Institutional grants from Institut Pasteur, INSERM and CNRS. DK is a scholar from the Pasteur Paris University International Doctoral Program funded by Institut Pasteur. MIG is supported by a grant from Agence National de Recherche (ANR) TRANSFACT (ANR-19-CE12-0007-02). MJ has been funded by an Allocation de Recherche Doctorale from La Ligue Nationale Contre le Cancer. MM was a scholar of the Pasteur Paris University International Doctoral Program and was supported by the Institut Pasteur, the European Union Horizon 2020 Research and Innovation Programme under the Marie Sklodowska-Curie grant agreement 665807 (COFUND-PASTEURDOC) and La Ligue Contre le Cancer - Allocation doctorale 4ème année de thèse. The authors are thankful to FAP and healthy participants for their generous engagement in the project, with special thanks to the members of Association Polyposes Familiales France for their engagement all along the project, to the ICAReB-Clin team for medical and technical support in the translational research program, to the Photonic Bioimaging UTechS for technical microscopy support and to Dr. C. Zimmer for scientific support.

Declaration of interests

The authors have no financial conflict of interest.

References

Aguera-Gonzalez, S., Burton, O. T., Vazquez-Chavez, E., Cuche, C., Herit, F., Bouchet, J., et al. (2017). Adenomatous polyposis coli defines Treg differentiation and anti-inflammatory function through microtubule-mediated NFAT localization. *Cell Reports, 21*, 181–194.

Bertrand, F., Muller, S., Roh, K. H., Laurent, C., Dupre, L., & Valitutti, S. (2013). An initial and rapid step of lytic granule secretion precedes microtubule organizing center polarization at the cytotoxic T lymphocyte/target cell synapse. *Proceedings of the National Academy of Sciences of the United States of America, 110*, 6073–6078.

Bradski, G. (2000). The OpenCV library. *Dr Dobb's Journal of Software Tools, 120*, 122–125.

Cuche, C., Mastrogiovanni, M., Juzans, M., Laude, H., Ungeheuer, M. N., Krentzel, D., et al. (2023). T cell migration and effector function differences in familial adenomatous polyposis patients with APC gene mutations. *Frontiers in Immunology, 14*, 1163466.

de Saint Basile, G., Menasche, G., & Fischer, A. (2010). Molecular mechanisms of biogenesis and exocytosis of cytotoxic granules. *Nature Reviews. Immunology, 10*, 568–579.

Dupre, L., Boztug, K., & Pfajfer, L. (2021). Actin dynamics at the T cell synapse as revealed by immune-related actinopathies. *Frontiers in Cell and Development Biology, 9*, 665519.

Harris, C. R., Millman, K. J., van der Valt, Gommers, R., Virtanen, P., Cournapeau, D., et al. (2020). Array programming with NumPy. *Nature, 585*, 357–362.

Hunter, D. J. (2007). Matplotlib: A 2D graphics environment. *Computing in Science & Engineering, 9*, 90–95.

Jacques, E., Buytaert, J., Wells, D. M., Lewandowski, M., Bennett, M. J., Dirckx, J., et al. (2013). MicroFilament analyzer, an image analysis tool for quantifying fibrillar orientation, reveals changes in microtubule organization during gravitropism. *The Plant Journal, 74*, 1045–1058.

Juzans, M., Cuche, C., Rose, T., Mastrogiovanni, M., Bochet, P., Di Bartolo, V., et al. (2020). Adenomatous polyposis coli modulates actin and microtubule cytoskeleton at the immunological synapse to tune CTL functions. *Immunohorizons, 4*, 363–381.

Kuhl, F. P., & Giardina, C. R. (1982). Elliptic Fourier features of a closed contour. *Computer Graphics and Image Processing, 18*, 236–258.

Lasserre, R., Charrin, S., Cuche, C., Danckaert, A., Thoulouze, M. I., de Chaumont, F., et al. (2010). Ezrin tunes T-cell activation by controlling Dlg1 and microtubule positioning at the immunological synapse. *The EMBO Journal, 29*, 2301–2314.

Mastrogiovanni, M., Di Bartolo, V., & Alcover, A. (2022). Cell polarity regulators, multifunctional organizers of lymphocyte activation and function. *Biomedical Journal, 45*, 299–309.

Mastrogiovanni, M., Juzans, M., Alcover, A., & Di Bartolo, V. (2020). Coordinating cytoskeleton and molecular traffic in T cell migration, activation, and effector functions. *Frontiers in Cell and Development Biology, 8*, 591348.

Mastrogiovanni, M., Vargas, P., Rose, T., Cuche, C., Esposito, E., Juzans, M., et al. (2022). The tumor suppressor adenomatous polyposis coli regulates T lymphocyte migration. *Science Advances, 8*, eabl5942.

Otsu, N. (1979). *A threshold selection method from gray-level histograms*. https://ieeexplore.ieee.org/document/4310076.

Puspoki, Z., Storath, M., Sage, D., & Unser, M. (2016). Transforms and operators for directional bioimage analysis: A survey. *Advances in Anatomy, Embryology, and Cell Biology, 219*, 69–93.

Randzavola, L. O., Strege, K., Juzans, M., Asano, Y., Stinchcombe, J. C., Gawden-Bone, C. M., et al. (2019). Loss of ARPC1B impairs cytotoxic T lymphocyte maintenance and cytolytic activity. *The Journal of Clinical Investigation*, *129*, 5600–5614.

Rezakhaniha, R., Agianniotis, A., Schrauwen, J. T., Griffa, A., Sage, D., Bouten, C. V., et al. (2012). Experimental investigation of collagen waviness and orientation in the arterial adventitia using confocal laser scanning microscopy. *Biomechanics and Modeling in Mechanobiology*, *11*, 461–473.

Ritter, A. T., Asano, Y., Stinchcombe, J. C., Dieckmann, N. M., Chen, B. C., Gawden-Bone, C., et al. (2015). Actin depletion initiates events leading to granule secretion at the immunological synapse. *Immunity*, *42*, 864–876.

Ritter, A. T., Kapnick, S. M., Murugesan, S., Schwartzberg, P. L., Griffiths, G. M., & Lippincott-Schwartz, J. (2017). Cortical actin recovery at the immunological synapse leads to termination of lytic granule secretion in cytotoxic T lymphocytes. *Proceedings of the National Academy of Sciences of the United States of America*, *114*, E6585–E6594.

Round, J. L., Humphries, L. A., Tomassian, T., Mittelstadt, P., Zhang, M., & Miceli, M. C. (2007). Scaffold protein Dlgh1 coordinates alternative p38 kinase activation, directing T cell receptor signals toward NFAT but not NF-kappaB transcription factors. *Nature Immunology*, *8*, 154–161.

Round, J. L., Tomassian, T., Zhang, M., Patel, V., Schoenberger, S. P., & Miceli, M. C. (2005). Dlgh1 coordinates actin polymerization, synaptic T cell receptor and lipid raft aggregation, and effector function in T cells. *The Journal of Experimental Medicine*, *201*, 419–430.

Sofroniew, M. B. (2009). Molecular dissection of reactive astrogliosis and glial scar formation. *Trends in Neurosciences*, *32*, 638–647.

Sokolov, P. A., Belousov, M. V., Bondarev, S. A., Zhouravleva, G. A., & Kasyanenko, N. A. (2017). FibrilJ: ImageJ plugin for fibrils' diameter and persistence length determination. *Computer Physics Communications*, *214*, 199–206.

Stinchcombe, J. C., Bossi, G., Booth, S., & Griffiths, G. M. (2001). The immunological synapse of CTL contains a secretory domain and membrane bridges. *Immunity*, *15*, 751–761.

Stinchcombe, J. C., Majorovits, E., Bossi, G., Fuller, S., & Griffiths, G. M. (2006). Centrosome polarization delivers secretory granules to the immunological synapse. *Nature*, *443*, 462–465.

Tamzalit, F., Tran, D., Jin, W., Boyko, V., Bazzi, H., Kepecs, A., et al. (2020). Centrioles control the capacity, but not the specificity, of cytotoxic T cell killing. *Proceedings of the National Academy of Sciences of the United States of America*, *117*, 4310–4319.

Tamzalit, F., Wang, M. S., Jin, W., Tello-Lafoz, M., Boyko, V., Heddleston, J. M., et al. (2019). Interfacial actin protrusions mechanically enhance killing by cytotoxic T cells. *Science Immunology*, *4*, eaav5445.

Tee, Y. H., Goh, W. J., Yong, X., Ong, H. T., Hu, J., Tay, I. Y. Y., et al. (2023). Actin polymerisation and crosslinking drive left-right asymmetry in single cell and cell collectives. *Nature Communications*, *14*, 776.

van der Walt, S., Nunez-Iglesias, J. L., Boulogne, F., Warner, J. D., Yager, N., et al. (2014). Scikit-image: Image processing in Python. *PeerJ*, *2*, e453.

Virtanen, P., Gommers, R., Oliphant, T. E., Haberland, M., Reddy, T., Cournapeau, D., et al. (2020). SciPy 1.0: Fundamental algorithms for scientific computing in Python. *Nature Methods*, *17*, 261–272.

CHAPTER FOUR

A comprehensive guide to study the immunological synapse using imaging flow cytometry

Andrea Michela Biolato[a,b], Liza Filali[a], Max Krecke[a,b], Clément Thomas[a], and Céline Hoffmann[a,c,*]

[a]Cytoskeleton and Cancer Progression, Department of Cancer Research, Luxembourg Institute of Health, Luxembourg, Luxembourg
[b]Faculty of Science, Technology and Medicine, University of Luxembourg, Esch-sur-Alzette, Luxembourg
[c]National Cytometry Platform, Luxembourg Institute of Health, Esch-sur-Alzette, Luxembourg
*Corresponding author: e-mail address: celine.hoffmann@lih.lu

Contents

1. Introduction	70
2. Materials	72
2.1 Cells	72
2.2 Cell culture	72
2.3 Cell labeling	73
2.4 Data acquisition and analysis	73
3. Methods	74
3.1 Cell culture	75
3.2 Target and effector cell preparation	76
3.3 Immunofluorescence staining: Extracellular staining	78
3.4 Co-culture and fixation	78
3.5 Immunofluorescence staining: Intracellular staining	78
3.6 Acquisition using imaging flow cytometry	79
3.7 Data analysis: IDEAS software	81
4. Concluding remarks	92
5. Notes	92
Acknowledgments	96
Bibliography	96

Abstract

Cytotoxic lymphocytes, such as cytotoxic T cells and natural killer (NK) cells, are instrumental in the recognition and eradication of pathogenic cells, notably those undergoing malignant transformation. Cytotoxic lymphocytes establish direct contact with cancer cells via the formation of a specialized cell-cell junction known as the lytic immunological synapse. This structure serves as a critical platform for lymphocytes to integrate surface signals from potential cancer cells and to direct their cytolytic

---

Methods in Cell Biology, Volume 193
ISSN 0091-679X
https://doi.org/10.1016/bs.mcb.2024.03.001

Copyright © 2025 Elsevier Inc.
All rights are reserved, including those
for text and data mining, AI training,
and similar technologies.

apparatus toward the confirmed targets. Conversely, cancer cells evolve synaptic defense strategies to evade lymphocyte cytotoxicity. This chapter delineates protocols using imaging flow cytometry to examine and quantify important subcellular processes occurring within cytotoxic lymphocytes and cancer cells engaged into an immunological synapse. These processes encompass the spatial redistribution of cytoskeletal components, vesicles, organelles and cell surface molecules. We specifically describe methods to generate and select conjugates between MDA-MB-231 breast cancer cells or K-562 leukemic cells and either the NK-92MI cell line or primary human NK cells. In addition, we detail procedures to evaluate the synaptic polarization of the actin cytoskeleton, CD63-positive vesicular compartments, MHC class I molecules, as well as the microtubule-organizing center in effector cells.

1. Introduction

The immunological synapse (IS) is an essential intercellular junction that enables communication between lymphocytes and antigen–presenting cells, including dendritic cells, macrophages and B cells. This junction is pivotal in orchestrating signaling cascades and supporting key cellular processes integral to immune responses, such as antigen recognition and presentation, and the regulation of lymphocyte activation, survival and differentiation. Additionally, it modulates immune responses through immune checkpoint pathways and targeted cytokine secretion. Moreover, the IS plays a central role in the selective destruction of diseased cells by cytotoxic lymphocytes, such as cytotoxic T lymphocytes (CTLs) and natural killer (NK) cells. Here, it serves as a platform for directing the release of lytic granules loaded with cytotoxic molecules towards target cells, enhancing killing efficacy while minimizing damage to nearby healthy cells. Although CTLs and NK cells are activated via different mechanisms, their direct cytotoxicity follows a series of similar key steps (Mace et al., 2014). These steps include the formation and maturation of a lytic IS, involving sequential remodeling of the actin cytoskeleton, the polarization of the microtubule–organizing center (MTOC) along with associated lytic granules toward the synaptic region, and the subsequent release of effector molecules, such as perforin, granzymes and granulysin, into the synaptic cleft. In response, target cells, notably cancer cells, have developed sophisticated synaptic defense strategies to escape lymphocyte cytotoxicity (Biolato et al., 2020; Cantoni et al., 2020; McKenzie & Valitutti, 2023; Ockfen et al., 2023). Recently, melanoma cells were shown to evade CTL attack by activating a synaptic membrane repair mechanism following perforin–induced perforation (Filali et al., 2022;

Khazen et al., 2016). This mechanism is characterized by a rapid, calcium-dependent relocation and exocytosis of lysosomes within melanoma cells at the synaptic area, leading to the restoration of membrane integrity and subsequent cell survival. Another example of synaptic defense strategy, which has been reported in breast cancer cells and chronic lymphocytic leukemia cells during their interaction with NK cells, involves the rapid accumulation of filamentous actin at the tumor cell side of the IS (Al Absi et al., 2018; Wurzer et al., 2021). Although the mechanistic details are still under investigation, such cytoskeletal remodeling is strongly associated with cancer cell resistance to NK cell-mediated cytotoxicity.

Further characterization of synaptic defense strategies requires experimental methods that enable the quantification and subcellular localization of multiple cell components within target cells. In addition, assessing the distribution of the actin cytoskeleton, MTOC and vesicles within immune effector cells stands a robust method for evaluating the functional state of an IS. In this regard, imaging flow cytometry (IFC) has emerged as a powerful tool to study the IS, enabling the imaging and analysis of a significant number of cell-cell conjugates, typically ranging from dozens to several hundreds, within a relatively short timeframe (Ahmed et al., 2009; Markey & Gartlan, 2019; Viswanath et al., 2017; Wabnitz et al., 2015; Zuba-Surma et al., 2007). By combining the high-throughput capacity of flow cytometry with the spatial resolution of microscopy, IFC allows both quantitative and qualitative analysis of processes occurring at both cell sides of the IS. Analysis of IFC images primarily involves delineating specific cellular regions of interest using "masks" and selecting "features" that align with the research question and the type of cellular component being investigated. Here, we present a detailed methodology for generating and imaging conjugates between MDA-MB-231 breast cancer cells or K-562 leukemic cells and either the NK-92MI cell line or primary human NK cells. We describe an optimized gating strategy designed for the selective identification of the conjugates suitable for IFC analysis. Additionally, we detail analytical strategies tailored to assess the degree of synaptic polarization of four different types of cellular components, including the actin cytoskeleton, CD63-positive vesicular compartments and MHC class I molecules—the most potent inhibitory ligands for NK cells—in target cells, as well as the microtubule-organizing center in effector cells. While this chapter provides specific examples for illustrative purposes, the methodologies detailed herein are versatile and can be adapted to address a broad spectrum of research questions.

2. Materials

2.1 Cells

- Effector cells: the NK-92MI cell line were purchased from ATCC (CRL-2408™). The primary NK cells were isolated from cryopreserved peripheral blood mononuclear cells (PBMCs; see Section 3.1) obtained from healthy, anonymous donor buffy coats provided by the Luxembourg Red Cross (Ref. project MAN-MAN-18-096).
- Target cells: the MDA-MB-231 breast adenocarcinoma cell line were purchased from ATCC (CRM-HTB-26™). The K-562 chronic myelogenous leukemia cell line were available at the Luxembourg Institute of Health and was authenticated through STR profiling analysis. Both cell lines were genetically modified to stably express the Emerald-Lifeact-7 (EmLA) F-actin reporter (Riedl et al., 2008) as previously described (Al Absi et al., 2018; Wurzer et al., 2021). The stable expression of an F-actin reporter facilitates accurate analysis of the actin cytoskeleton in target cells, effectively eliminating the risk of cross-contamination from signal originating from the effector cell. For methods to visualize the actin cytoskeleton without genetic manipulation, refer to Note 1.

2.2 Cell culture

- High glucose Dulbecco's modified Eagle's medium (DMEM), GlutaMAX™ Supplement (#10566016, Thermo Fisher Scientific).
- Roswell Park Memorial Institute (RPMI)-1640 medium (#61870010, Thermo Fisher Scientific).
- Fetal Bovine Serum (FBS), heat-inactivated (#10270-106, Gibco™).
- Horse Serum (HS), heat-inactivated, New Zealand Origin (JM) (#10368902, Gibco™).
- Penicillin/Streptomycin mixture 10,000 units/mL (#15140122 Gibco™).
- Sodium pyruvate (#11360.070 Gibco™).
- HEPES (#156030-080 Gibco™).
- MEM-NEA (#11140-035 Gibco™).
- Puromycin dihydrochloride (#sc-108071, Santa Cruz Biotechnology).
- Phosphate-buffered saline 1× (PBS, without $CaCl_2$ and $MgCl_2$, pH 7.4) (#14040091, Life Technologies).
- 0.05% (w/v) Trypsin-EDTA solution (#25300062, Sigma-Aldrich).
- Trypan blue solution, 0.4% (#15250061, Gibco™).

Exploring the immunological synapse: Insights from imaging flow cytometry 73

- Neubauer cell counting hemacytometer (#15900406, Fisher Scientific).
- Cell culture incubator allowing standard cell culture conditions in a humidified atmosphere (37 °C, 5% CO_2); cell culture laminar flow hood and an inverted microscope for cell counting.
- Cell culture flasks 25 cm^2 or 75 cm^2 (CELLSTAR$^®$ #658175/#690175, Greiner Bio-One).
- CELLSTAR$^®$ round-bottom polypropylene test tubes with cap, 15 or 50 mL (#188271/#227261, Greiner Bio-One).
- Micro-centrifuge tubes 1.5 mL (#11926955, Fisher Scientific).
- Serological pipettes (CELLSTAR$^®$ 2, 10 and 25 mL, Greiner Bio-One).
- Centrifuge Thermo Scientific™ Sorvall™ ST40 (Fisher Scientific).
- Bench top MiniSpin Plus centrifuge (Eppendorf).
- NK cell isolation kit, human (#130-092-657, Miltenyi).
- Human IL-2 IS, premium grade (#130-097-746, Miltenyi).
- Human IL-15, premium grade (#130-095-764, Miltenyi).
- LS column (#130-042-401, Miltenyi).
- QuadroMACS separator (#130-090-976, Miltenyi).

2.3 Cell labeling

- AutoMACS$^®$ Buffer (#130-091-376, Miltenyi Biotec).
- Paraformaldehyde 16% w/v, methanol free (#11400580, Thermo Fisher Scientific).
- Triton X-100 (#X100-100ML, Sigma–Aldrich).
- CellTrace™ Violet (CTV) proliferation kit (#C34557, Thermo Fisher Scientific).
- Zombie NIR™ fixable viability kit (#423106, BioLegend).
- PE/DAZZLE™ 594 anti-human HLA-A,B,C antibody (clone: W6/32, #311440, BioLegend).
- PE anti-human CD63 antibody (clone: H5C6, # 353004, BioLegend).
- Alexa Fluor$^®$ 594 anti-Y-tubulin (stain for the MTOC) (clone: D-10, #sc-17,788, Santa Cruz Biotechnology).

2.4 Data acquisition and analysis

- ImageStreamX Mark II (Amnis$^®$, Cytek Biosciences) imaging flow cytometer with five built-in lasers (405, 488, 561, 640, and 785 nm).
- Ideas software (version 6.2.64.0, Cytek Biosciences).
- GraphPad Prism software (version 10.0, GraphPad software).

3. Methods

Fig. 1 illustrates the analytical strategies, notably the specific combinations of masks and features, proposed for evaluating synaptic polarization of four types of cellular component, including surface molecules, cytoplasmic

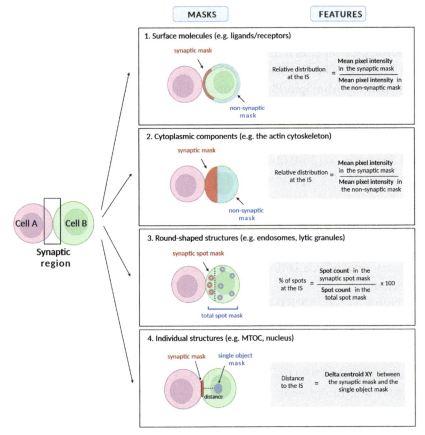

Fig. 1 Overview of mask and feature combinations for quantifying synaptic polarization of different types of cellular components. Cells engaged in a conjugate are depicted in pink and green. The distinction between effector and target cells is not specified, as the analysis is equally applicable to both cell types within the conjugate. 1. Masks and features used for plasma membrane proteins, such as inhibitory/activating ligands on the target cell surface and inhibitory/activating receptors on the effector cell surface. 2. Masks and features used for cytoplasmic components, including the actin cytoskeleton and cytosolic proteins. 3. Masks and features used for round-shaped structures, like vesicles and lytic granules. 4. Masks and features used for individual cell components, such as the nucleus and the MTOC. *Created with BioRender.com.*

components such as the actin cytoskeleton, round-shaped structures including vesicles, and individual structures such the nucleus and the MTOC. These strategies are versatile and can be applied to various cellular components, as well as to both effector and target cells, thereby enabling the exploration of a diverse range of research questions.

In the following sections, we comprehensively detail the protocols for generating and staining conjugates between NK cells (using either the NK-92MI cell line or primary NK cells obtained from healthy donors) and cancer cells (specifically the MDA-MB-231 breast adenocarcinoma cell line or the K-562 chronic myelogenous leukemia cell line). We outline the procedures for acquiring and analyzing IFC data, with reference to the strategies presented in Fig. 1. While the strategy is primarily designed for IFC, it is important to mention that the masking techniques it employs can be adapted and utilized in various other software platforms tailored for microscopy analysis, such as ImageJ or Icy software. To illustrate the practical application of these methods, we provide specific examples where we analyze the synaptic polarization of four different cellular components, including (1) HLA-A,-B,-C on the surface of MDA-MB-231 cells conjugated with NK-92MI cells; (2) filamentous actin in MDA-MB-231 and K-562 cells conjugated with NK-92MI cells; (3) CD63-positive vesicles in MDA-MB-231 cells, whether conjugated with NK-92MI or primary NK cells; and (4) the MTOC in NK-92MI cells when conjugated with either MDA-MB-231 or K-562 cells. It should be noted that both cancer cell lines have been modified to express the fluorescent actin reporter EmLA (Al Absi et al., 2018; Wurzer et al., 2021), facilitating identification of target cells, as well as visualization of the actin cytoskeleton without additional staining.

3.1 Cell culture

Cells are maintained in a CO_2 incubator upon standard conditions (37 °C, 5% CO_2).

- MDA-MB-231 cells are cultured in DMEM supplemented with 10% FBS and 1% Penicillin/Streptomycin. Cell passaging is carried out twice weekly, initiated specifically when the cell confluence reaches between 80% and 90%. Briefly, cells are washed with PBS, detached using Trypsin for 3–5 min, and subsequently transferred to a new culture vessel with a dilution factor of 1:4. Cells used in this protocol do not exceed passage number 10.

- K-562 cells are cultured in RPMI medium supplemented with 10% FBS and 1% Penicillin/Streptomycin. They are maintained in 75 cm^2 flasks and undergo bi-weekly subculturing to ensure a cell concentration range of 3×10^5 to 1×10^6 cells/mL.
- NK-92MI cells are cultured in RPMI medium supplemented with 10% FBS, 10% HS and 1% Penicillin/Streptomycin. They are maintained within 75 cm^2 flasks and undergo bi-weekly subculturing, to ensure a cell concentration range of 3×10^5 to 1×10^6 cells/mL.
- Human primary NK cells are isolated by negative selection 1 day prior to the experiment and are maintained as described below.

  1. Prepare RPMI medium supplemented with 10% FBS, 1% HEPES 10 mM, 1% MEM-NEA, 1% sodium pyruvate and 1% Penicillin/Streptomycin.
  2. Thaw cryopreserved PBMCs in 37 °C water bath for few minutes.
  3. Transfer the cell suspension in pre-heated medium in a 15 mL falcon tube (dilution 1:10 of the cell suspension volume).
  4. Centrifuge the cells at 300 × g for 5 min at room temperature (RT).
  5. Wash the cells in 4 °C MACS buffer.
  6. From this point forward, utilize the human NK cell isolation kit as per the manufacturer's instructions (#130-092-657, Miltenyi). Briefly, start by counting the PBMCs and resuspend them in the recommended volume of buffer along with the biotin-antibody cocktail. Incubate 5 min at 4 °C. Then, add the microbeads cocktail to the mixture and incubate for an additional 10 min at 4 °C. Following this, proceed with the manual cell separation using the appropriate column and magnet.
  7. After isolation, wash primary NK cells with 15 mL PBS.
  8. Count the cells using a 0.4% Trypan blue solution and a Neubauer counting chamber.
  9. Resuspend the cells in pre-heated medium supplemented with IL-2 (100 IU/mL) and IL-15 (10 ng/mL), achieving a final cell concentration of 2×10^6 cells/mL.
  10. Incubate the cells overnight prior to utilizing them in the experiment (Note 2).

3.2 Target and effector cell preparation

As a reminder, the target cells used in this protocol express the Emerald-LifeAct protein, an actin reporter, which eliminates the need for additional staining to identify target cells. However, should wild-type cells be employed, a labeling step can be introduced at this stage using a fluorescent dye for cytoplasmic staining, ensuring compatibility with the rest of the fluorescent panel (Note 3).

3.2.1 Target cells: MDA-MB-231 and K-562 cells

For non–adherent K-562 cells, proceed directly to step 6. For MDA-MB-231:

1. Once the cells attain 80–90% confluency, aspirate the culture medium and gently wash the adherent cells with PBS.
2. Discard the PBS and detach the cells using a 0.05% (w/v) Trypsin-EDTA solution for 3–5 min at 37 °C (Note 4).
3. Add complete medium to the cells and transfer the cell solution into a 15 mL falcon tube.
4. Centrifuge the cells at $300 \times g$ for 5 min at RT.
5. Resuspend the cell pellet in MACS buffer for 5 min. This step will enhance disaggregation of cell clusters.
6. Count the cells using a 0.4% Trypan blue solution in a Neubauer counting chamber.
7. Centrifuge the cell suspension at $300 \times g$ for 5 min at RT.
8. Resuspend 3×10^5 cells in 100 μL of complete medium within a 1.5 mL micro-centrifuge tube.
9. Allow a recovery period of 20 min for the cells prior to initiating the co-culture.

Cells are then ready for conjugation or may undergo further extracellular staining procedures (for details, refer to Section 3.3). Set aside an aliquot of cells (3×10^5 cells in 35 μL PBS) as a single-stain control to calculate the compensation matrix for the Emerald fluorescence channel prior sample acquisition.

3.2.2 Effector cells: NK-92MI and human primary NK cells

1. Count the cells using 0.4% Trypan blue solution.
2. Stain 1×10^6 cells with CTV (1:8000) and Zombie-NIR (1:2000) in 1 mL of PBS for 25 min at RT (Note 5). In parallel, prepare a single-stain tube (either CTV or Zombie-NIR only) to create the compensation matrix before starting the acquisition on the cytometer.
3. Wash the cells twice by centrifugation at $500 \times g$ for 3 min using complete medium.

4. Resuspend cells at 6×10^6/mL in appropriate complete medium, and transfer 100 µL, into a 1.5 mL micro-centrifuge tube.
5. Allow a recovery period of 15 min for the cells prior to initiating the co-culture.

3.3 Immunofluorescence staining: Extracellular staining

1. Incubate 1×10^6 target cells for 30 min at 4 °C in 100 µL of MACS buffer containing fluorophore-labeled antibodies, e.g., PE/DAZZLE™ 594 anti-HLA-A,-B,-C antibodies at a 1:40 dilution. Simultaneously, prepare a mono-color tube of unconjugated cells for channel compensation (Note 6).
2. Wash the cells using MACS buffer and centrifuge at $500 \times g$ for 3 min.
3. Resuspend cells to reach 3×10^5 cells in 100 µL of appropriate complete medium within a 1.5 mL micro-centrifuge tube.
4. Allow a recovery period of 15 min at 37 °C for the cells prior to initiating the co-culture.

3.4 Co-culture and fixation

1. In a 1.5 mL micro-centrifuge tube, gently mix 6×10^5 NK-92MI cells in 100 µL with 3×10^5 MDA-MB-231 cells also in 100 µL, achieving a cell-to-cell ratio of 2:1.
2. Incubate the cell co-culture at 37 °C for 40 min.
3. Post-incubation, carefully remove 100 µL of the medium and replace it with 100 µL of 4% PFA, resulting in a final concentration of 2% PFA (Note 7). Continue incubation at 37 °C for an additional 15 min.
4. Pellet the cells through centrifugation at $500 \times g$ for 3 min.
5. Wash the cells using PBS.
6. If no further labeling is needed, resuspend the cells in 35 µL of PBS. Subsequently store the cell suspension at 4 °C until IFC acquisition.

3.5 Immunofluorescence staining: Intracellular staining

1. Permeabilize the cells with 0.1% Triton X-100 for 5 min at RT.
2. Subsequent to a gentle wash in MACS buffer, resuspend the cell pellet in 100 µL of MACS buffer containing fluorophore-labeled antibodies, e.g., PE anti-CD63 antibody at a 1:40 dilution, or AF594-anti-Y-tubulin antibody at a 1:20 dilution. In parallel, prepare a single-stain tube to create the compensation matrix.
3. Incubate the cells for 2 h at RT in the dark.

Exploring the immunological synapse: Insights from imaging flow cytometry   79

4. Perform a double wash of the cells by centrifuging at $500 \times g$ for 3 min using PBS.
5. Resuspend the cells in 35 µL of PBS and store them at 4 °C until IFC acquisition.

3.6 Acquisition using imaging flow cytometry

The image acquisition procedure outlined below is performed using the ImageStreamX Mark II equipped with five integrated lasers (405, 488, 561, 640, and 785 nm). Data acquisition is conducted using the INSPIRE software. The acquired data are stored as raw image files (.rif format) including both the instrument settings and the actual image data.

3.6.1 Instrument setup

1. Start up the ImageStreamX and perform calibration using the Assist feature as per manufacturer's instructions.
2. Load a sample, create a "Raw Max Pixel" histogram for each channel. Adjust the laser power to achieve intensity values between 100 and 4000 counts. This ranges ensures coverage of the dynamic range while preventing saturation. Based on the staining protocol described in this chapter, the following laser powers were applied: 405 nm (30 mW), 488 nm (40 mW), 561 nm (100 mW), and 642 nm (150 mW).
3. Imaging was performed with a 60× magnification objective, using low-speed and high-sensitivity settings for optimal imaging.
4. Select the extended depth of field (EDF) mode if appropriate. This mode is particularly recommended for spot detection and was employed for samples stained for CD63 or Y-tubulin; See Note 8 and Note 9 for additional information.
5. Load single-stained samples to construct the compensation matrix using the compensation wizard tool. It is necessary to set the matrix before sample acquisition to enable accurate gating on conjugates, as they are identified based on the violet signal from CTV and the green signal from EmLA.

3.6.2 Data acquisition

1. Gently resuspend the cell pellet and load the sample tube into the cytometer.
2. Establish a gating strategy to identify cells of interest for acquisition:
   Create a scatterplot of area versus aspect ratio of bright field channel to gate on cells and exclude beads and debris (Fig. 2A).

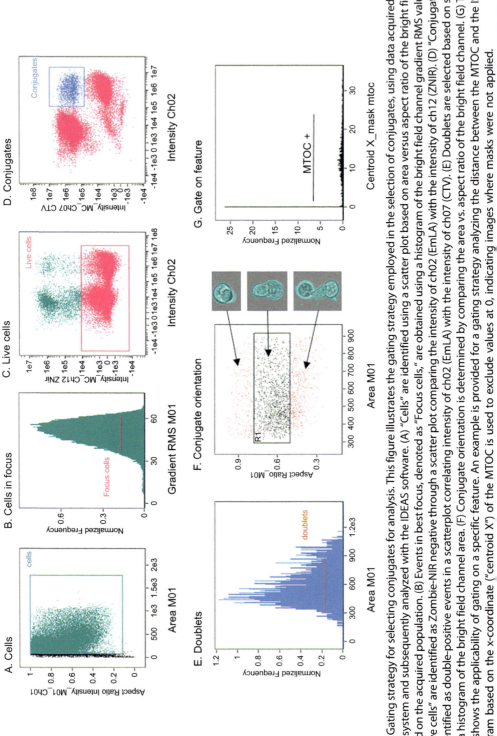

Fig. 2 Gating strategy for selecting conjugates for analysis. This figure illustrates the gating strategy employed in the selection of conjugates, using data acquired via an IFC system and subsequently analyzed with the IDEAS software. (A) "Cells" are identified using a scatter plot based on area versus aspect ratio of the bright field, applied on the acquired population. (B) Events in best focus, denoted as "Focus cells," are obtained using a histogram of the bright field channel gradient RMS values. (C) "Live cells" are identified as Zombie-NIR negative through a scatter plot comparing the intensity of ch02 (EmLA) with the intensity of ch12 (ZNIR). (D) "Conjugates" are identified as double-positive events in a scatterplot correlating intensity of ch02 (EmLA) with the intensity of ch07 (CTV). (E) Doublets are selected based on size, using a histogram of the bright field channel area. (F) Conjugate orientation is determined by comparing the area vs. aspect ratio of the bright field channel. (G) This panel shows the applicability of gating on a specific feature. An example is provided for a gating strategy analyzing the distance between the MTOC and the IS. A histogram based on the x-coordinate ("centroid X") of the MTOC is used to exclude values at 0, indicating images where masks were not applied.

Create a scatterplot correlating the intensity of ch02 (EmLA) against the intensity of ch07 (CTV) for identifying double-positive events, referred to as 'conjugates' (Fig. 2D).

3. Set the file acquisition parameters to collect 5×10^3 to 8×10^3 events of conjugates per sample (Note 10).

4. Save the settings as a template after the first experiment, it will be named with the suffix .ist. Load the template for subsequent experiments to maintain consistency.

3.7 Data analysis: IDEAS software

3.7.1 Compensation and data processing

1. Generate a compensation matrix file (.ctm) using the "Create New Matrix" wizard within the IDEAS software. Upload the .rif files of the single-stained samples that are labeled as "noBF" (no Bright field). Follow the instructions provided in the compensation dropdown menu to create the .ctm file. See Note 11 for additional information.

2. Proceed to open the .rif file corresponding to the fully stained sample and apply the previously generated .ctm file. This action will automatically generate both .cif (compensated image file) and .daf (data analysis file) files.

3.7.2 Gating strategy

Different gates are established within the analysis area to exclude undesirable events.

1. Gate on "cells" by creating a scatterplot of area versus aspect ratio of bright field channel (Fig. 2A). Exclude objects with small areas, such as beads and debris, by applying a size-based selection.

2. Gate on "focus cells" using a histogram of the bright field channel gradient RMS values based on the population selected as cells in the previous step (Fig. 2B). See Note 12.

3. Create a gate for "live cells" by implementing a scatter plot comparing the intensity of ch02-EmLA against the intensity of ch12-ZNIR. Identify live cells as those unstained by Zombie-NIR (Fig. 2C).

4. To identify "conjugates," generate a scatter plot within the analysis area comparing the intensity of ch02-EmLA against the intensity of ch07-CTV. Delineate a region encompassing the double-positive events to categorize them as "conjugates" (Fig. 2D). This population should contain at least one target cell (EmLA) and one NK cell (CTV).

5. Optional: Create a gate for "doublets" using a histogram based on the area of the bright field channel. In our samples, doublets typically fall within the size range of 300–900 μm^2. This gate filters out objects with smaller areas, such as cut cells, and larger ones, such as conjugates with more than two cells or aggregates (Fig. 2E). This is particularly relevant when working with MDA-MB-231 cells or other "sticky" cells.

6. Gate on "oriented doublets" by adding a scatter plot based on the area versus the aspect ratio of the bright field channel (Fig. 2F). A high aspect ratio indicates conjugates in a tilted position, whereas a low ratio includes conjugates with more than two cells. The gating at this step relies on the same parameters as in Fig. 2A. However, it is easier to verify orientation and establish the boundaries of the gate solely on conjugates rather than on the entire population.

7. Gate on one specific feature after the relevant masks and features have been calculated, as detailed in the following section (see Sections 3.7.3 and 3.7.4). For instance, in Fig. 2G, we employed a histogram based on the feature "Centroid X" to determine the x-coordinate of the MTOC. In this histogram, values at 0 correspond to images where no mask were applied, thus excluding images where the MTOC was not detected.

This gating strategy may still include conjugates comprising more than two cells, as well as cell doublets oriented in a manner that renders the immune synapse region challenging to analyze. Hence, we advise to undertake a manual inspection of the images. This additional control aims to exclude conjugates that are unsuitable for analysis, particularly those exhibiting misorientation or incorrectly positioned masks.

3.7.3 Region of interest definition (Mask)

Masks refers to designated regions of interest (ROIs) within an image. These masks are used to delineate individual cells or distinct subcellular regions or components, allowing for a focused analysis (Dominical, Samsel, & McCoy, 2017). Within these masks, a variety of measurements or "features" can be quantified, including fluorescence intensity, shape metrics and location (xy-coordinates), refer to Section 3.7.4 for further details. In the present section, we detail the process of creating masks specifically designed to identify various relevant entities such as conjugates, individual cell within a conjugate, the synaptic region or the rest of the cell, round-shaped cytoplasmic components, like vesicles and organelles, and individual intracellular structures, notably the nucleus and the MTOC. For practical illustration, we

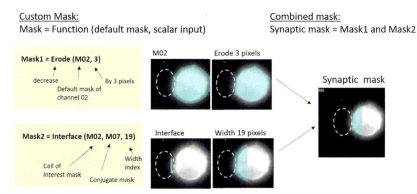

Fig. 3 Mask definition: example of the Synaptic mask creation. On the left side, detailed descriptions of the custom masks utilized for synaptic mask creation are provided. Mask1 and Mask2 are based on the "Erode" and "Interface" functions, respectively. Images represent a conjugate between a target cell (MDA-MB-231, EmLA channel displayed in gray) and a NK-92MI cell, with the position delineated by dashes. The masks are highlighted in light blue within the images. On the right side, the final combined mask is presented.

present examples pertinent to our research focus (Fig. 3). These examples include the analysis of synaptic polarization of (A) surface molecules such as HLA-A,-B,-C molecules on the target cell surface (which can act as potent inhibitory ligands for NK cells (Quatrini et al., 2021; Sivori et al., 2020)), (B) the actin cytoskeleton within the target cell (which correlates with increased resistance to NK cell-mediated cytotoxicity (Al Absi et al., 2018; Wurzer et al., 2021)), (C) multivesicular bodies as assessed by CD63 staining within the target cell, and (D) the MTOC position within NK cells (a marker for the activation of effector functions). See Note 13.

3.7.3.1 Mask definition

There are three types of masks: (1) Default masks are automatically generated and cannot be modified; they are named from M01 to M12, corresponding to the 12 channels. (2) Custom masks are generated from various functions, with 14 available options including Dilate, Erode, Interface, Object and Threshold. (3) Combined masks are created using Boolean logic to combine and subtract masks with "And/Or/Not" operations.

To create a new mask, open the mask manager window and choose a function from the dropdown menu. Fig. 3 outlines the process for creating the "Synaptic mask." Masks are named according to the function, followed by the default mask (channel of interest) and the scalar input for adjustment, enclosed in brackets. For instance, mask1 is based on the "Erode" function that removes the selected number of pixels (here in the example 3) from all edges of the selected mask (Default mask from channel2). The second mask relies on the "Interface" function, which identifies pixels within an object where the object comes into contact with a second object, specifically ch02 and ch07 in this instance. Subsequently, the mask is expanded using a width index that determines the breadth of the interface mask from the contact point toward the cell of interest. In our analysis, we utilize 19 pixels to encompass one-third of the target cell. For more detailed information, please refer to Note 14. The masks are then combined using an "and" operation; therefore, the synaptic mask corresponds to the overlap of Mask1 and Mask2.

3.7.3.2 Surface molecules: Synaptic polarization of HLA-A,B,C within target cells

The masks employed for such analysis selectively isolate the periphery/plasma membrane of the target cell, excluding the cytoplasm by using a combination of dilate and erode function (Fig. 4A). The "synaptic mask" is designed to encompass one-third of the target cell's periphery by using the interface function extended of 19 pixels, focusing on the segment of the plasma membrane in close proximity to the effector cell. Conversely, the "non-synaptic mask" is applied to cover the plasma membrane spanning the remaining portion of the cell.

- Synaptic mask: Dilate(M02,0) And Not Erode(M02,10) And Interface (M02, M07, 19). See Note 15 and Note 16.
- Non-synaptic mask: Dilate(M02, 0) And Not Erode(M02, 10) And Not Interface(M02, M06, 19).

3.7.3.3 Cytoplasmic components: Synaptic polarization of the actin cytoskeleton within target cells

The "synaptic mask" is designed to encompass the synaptic area within the target cell expanding into the cell interior and covering approximately one-third of the cell's total area, as depicted in Fig. 4B. The corresponding "non-synaptic mask" is applied to the remaining two thirds of the cell, this area being most distal from the effector cells and not actively engaged in the synaptic interaction.

- Synaptic mask: Erode(M02,3) And Interface(M02,M07,19).
- Non-synaptic mask: Erode(M02,3) And Not Interface(M02,M07,19).

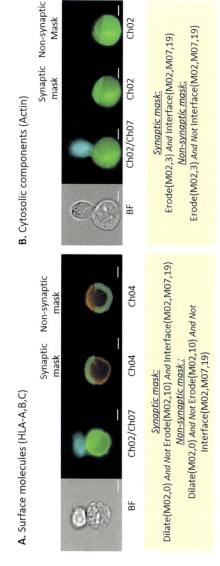
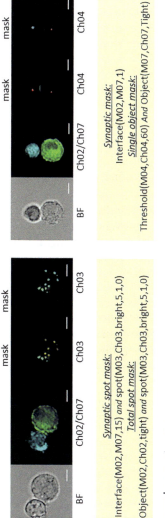
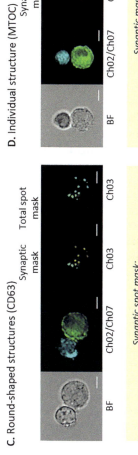

Fig. 4 See figure legend on next page.

3.7.3.4 Round-shaped structures: Synaptic polarization of CD63-positive vesicles within target cells

The synaptic mask is crafted to specifically cover CD63-positive vesicles (referred to as "spots") located within the synaptic region of the target cell, i.e. the first third of cell's total area that is proximal to the effector cell (Note 17). A second mask is designed to encompass CD63-positive vesicles distributed throughout the entire cell (including the synaptic and non-synaptic regions) (Fig. 4C).

- Synaptic spot mask: Interface(M02,M07,15) and spot(M03,Ch03, bright,5,1,0).
- Total spot mask: Object(M02,Ch02,tight) and spot(M03,Ch03, bright,5,1,0).

Various inputs are utilized to define the "spot" function. In this case, we employed the characteristics of "bright" with a spot-to-cell background ratio of 5, along with maximum and minimum radii ranging between 1 and 0.

3.7.3.5 Individual structure: Synaptic polarization of the MTOC within the effector cell

The single object mask is designed to selectively capture the MTOC of the CTV-stained NK cell and exclude the MTOC of the target cell, ensuring that data from the target cell MTOC does not confound the result. The threshold function is used to detect the bright MTOC spot and combine to the object function applied to the CTV channel. The XY-coordinates of the centroid of single object mask are compared with the XY-coordinates of the centroid of a synaptic mask specifically designed to cover the IS interface. Importantly, this synaptic mask is not expanded to ensure more

Fig. 4 Imaging flow cytometry (IFC) panels showcasing different types of masks (in light blue). A bright field image of conjugates between NK-92MI cells and target cells (MDA-MB-231) is shown on the left of each figure panel. NK cells are labeled using CTV and appears in cyan (ch07-CTV) while the target cells appear green due to the expression of the actin reporter Emerald-LifeAct (ch02–EmLA). The masks for each panel is defined below. (A) HLA-A,-B,-C on the target cell surface: the synaptic mask encompasses the plasma membrane in the IS region, while the non-synaptic mask covers the plasma membrane of the remaining cell area. (B) Actin cytoskeleton within the target cell: the synaptic mask covers the cytoplasm within the IS region. The non-synaptic mask conversely extends over the cytoplasm in the rest of the cell. (C) CD63 labeled vesicles: the synaptic spot mask is designed to detect CD63 vesicles within the synaptic region. The total spot mask, on the other hand, identifies CD63 vesicles throughout the entire cell. (D) MTOC: the synaptic mask is employed to detect the interface region between the two cells, while the single object mask is used to identify the MTOC within the NK cell. Scale bar $= 7 \mu m$.

accurate calculation of the centroid; the value is set at 1 pixel instead of 19 previously used to cover one third of the cell area (Fig. 4D).

- Single object mask: Threshold(M04,Ch04,60) And Object(M07,Ch07, Tight).
- Synaptic mask: Interface(M02,M07,1).

3.7.4 Features extraction and calculation

Different features are tailored to the analysis of the distinct types of subcellular components studied. The "Mean pixel intensity" feature is utilized to evaluate the synaptic polarization of surface molecules, such as HLA-A,-B,-C, as well as the actin cytoskeleton. The "spot count" feature is applied for assessing the synaptic polarization of round-shaped structures, such as CD63-positive vesicles. Lastly, the "delta centroid XY" feature is employed to evaluate the synaptic polarization of individual cellular components, such as the MTOC. This feature enables quantification of the distance between the MTOC and the synaptic interface. Features are generated by applying a feature from the default list provided by the IDEAS software, e.g. mean pixel intensity, to a specific mask. In addition, there is the possibility to combine two features to perform a mathematical operation using data extracted from two different masks. The following section provides further details on the methodologies for applying features in the analysis of the IS, as well as the calculations that can be conducted, drawing upon the examples outlined in this chapter.

3.7.4.1 Surface molecules: Synaptic polarization of HLA-A,B,C within target cells

In the example illustrated in Fig. 5A, MDA-MB-231 cells were first pre-stained for surface HLA-A,-B,-C molecules and then co-cultured with NK-92MI cells. After applying relevant masks (as shown in Fig. 4A), the synaptic polarization of HLA-A,-B,-C molecules was evaluated by calculating the ratio of "mean fluorescence intensity" within the synaptic mask to that of the non-synaptic mask.

- Generate the HLA-A,-B,-C fluorescence intensity feature for both ROIs by applying the "mean pixel intensity" feature to the synaptic and the non-synaptic masks:

    "Mean Pixel_synaptic mask_Ch04-PE/DAZZLE™ 594" and "Mean Pixel_non-synaptic mask_Ch04-PE/DAZZLE™ 594."
- Combine the newly created features to calculate the relative distribution of HLA-A,-B,-C at the IS:

    "Mean Pixel_Mask IS_Ch04-PE/DAZZLE™ 594"/"Mean Pixel_Mask ligand non-IS_Ch04-PE/DAZZLE™ 594."

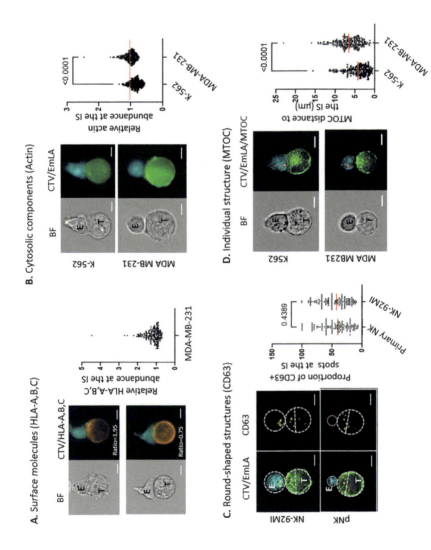

Fig. 5 See figure legend on opposite page.

In our observations, we noted that HLA-A,-B,-C molecules are not evenly distributed across the plasma membrane of MDA-MB-231 cells interacting with NK cells, with some cells displaying a pronounced accumulation of HLA-A,-B,-C molecules at the IS (ratio = 1.95), while others did not (ratio = 0.75) (Fig. 5A) (Al Absi et al., 2018).

3.7.4.2 Cytoplasmic components: Synaptic polarization of the actin cytoskeleton within target cells

Fig. 5B depicts a comparative analysis of the synaptic polarization of the actin cytoskeleton within K-562 and MDA-MB-231 cells upon their conjugation with NK-92MI cells. The polarization of actin at the target cell side of the IS was quantified by calculating the ratio of the mean fluorescence intensity within the synaptic mask relative to that within the non-synaptic mask. The criteria for these masks are delineated in Fig. 4B.

- Generate the actin fluorescence intensity feature in both ROIs by applying the "mean pixel intensity" feature to the synaptic and the non-synaptic masks:

    "Mean Pixel_synaptic mask_Ch02-EmLA" and "Mean Pixel_non-synaptic mask_Ch02-EmLA."

- Combine the newly created features to calculate the relative distribution of actin at the IS:

Fig. 5 Illustrative analysis of subcellular component distribution and synaptic polarization within conjugates of NK cell and cancer cell. (A) Relative abundance of HLA-A,-B,-C molecules at the IS in MDA-MB-231 cells conjugated with NK-92MI cells. Bright field images on the left display two distinct conjugates. NK cells (E) are labeled in cyan and HLA-A,-B,-C molecules are depicted in orange in the MDA-MB-231 cells (T). Top panels illustrate increased levels of HLA-A,-B,-C at the IS in one MDA-MB-231 cell, while bottom panels show uniform distribution of these molecules in another MDA-MB-231 cell. The chart represents data from 189 conjugates. (B) F-actin distribution in MDA-MB-231 vs. K-562 cells conjugated with NK-92MI cells. The left panels show examples of F-actin (green; EmLA actin reporter) distribution in MDA-MB-231 and K-562 cells interacting with NK-92MI cells. The chart summarizes data from 266 conjugates. The red dash line indicates a ratio value of 1. (C) CD63-positive vesicle abundance at the IS in MDA-MB-231 cells conjugated with NK-92MI vs. primary NK cells. Examples of CD63 vesicles (yellow) in MDA-MB-231 interacting with NK cell types are shown in the left panels. The chart compiles data from 137 conjugates. (D) MTOC positioning in NK-92MI cells conjugated with K-562 cells vs. MDA-MB-231 cells. The left panels display MTOC locations (red) in interactions between NK-92MI cells and K-562 cells or MDA-MB-231 cells. The chart shows MTOC distance from the IS centroid in these interactions, based on 214 conjugates. A Mann-Witney test was employed to assess statistical significance.

"Mean Pixel_synaptic mask_Ch02-EmLA"/"Mean Pixel_non-synaptic mask_Ch02-EmLA."

The data demonstrates a markedly higher incidence of synaptic polarization of the actin cytoskeleton in MDA-MB-231 cells relative to K-562 cells (Fig. 5B). This observation aligns with the more resistant phenotype of MDA-MB-231 cells and the significant susceptibility of K-562 cells, commonly used as a standard reference in evaluating NK cell cytotoxicity (Al Absi et al., 2018).

3.7.4.3 Round-shaped structures: Synaptic polarization of CD63-positive vesicles within target cells

Fig. 4C presents a comparative evaluation of the synaptic polarization of CD63-positive vesicles in MDA-MB-231 cells following their conjugation with either NK-92MI cells or primary NK cells derived from a healthy donor. This analysis quantified the proportion of CD63-positive labeled vesicles localized within the synaptic region. The quantification is achieved by employing the "spot count" feature on the synaptic spot and total spot masks described in Fig. 4C.

- Quantify the number of spots in both ROIs by applying the "spot count" feature. Apply this feature to both the synaptic mask ("Spot Count_synaptic spot mask_4") and the entire cell mask ("Spot Count_total spot mask_4").
- Calculate the percentage of spots at the IS using the following formula: "(Spot Count_vesicles IS_4/Spot Count_vesicles tot_4) × 100."

The findings reveal that the subcellular distribution of CD63-positive vesicles within the MDA-MB-231 cells demonstrate notable variability among individual cells (Fig. 5C). Some cells display a pronounced accumulation of these vesicles at the IS, while others show only a minimal presence. Intriguingly, no statistically differences were observed in vesicular distribution when target cells were analyzed in interaction with either NK-92MI or primary NK cells.

3.7.4.4 Individual structure: Synaptic polarization of the MTOC within the effector cell

In this analysis, NK-92MI cells were conjugated with either K-562 or MDA-MB-231 cells (Fig. 5D). The polarization of the MTOC toward the IS (a key metric in evaluating NK cell activation) was evaluated by measuring the distance between the centroid of the MTOC mask and that of the IS interface (Fig. 4).

- Use the X-centroid feature on the single object mask. This feature is specifically designed to facilitate the exclusion of values equal to 0, which indicate a lack of MTOC detection, aligning with the gating strategy shown in Fig. 2G.
- Calculate the distance between two masks using the "XY delta centroid" feature:

  "Delta Centroid XY_single object mask_Ch04_IntensityWeighted_synaptic mask_Ch01."

The results reveal a pronounced synaptic polarization of the MTOC in NK-92MI cells when conjugated with K-562 cells, as evidenced by a shorter average distance between the MTOC and the IS, recorded at $4.2 \pm 2\,\mu m$. Conversely, when NK-92MI cells are conjugated with MDA-MB-231 cells, the average distance of the MTOC to the IS is significantly greater, measuring $6.5 \pm 3\,\mu m$ (Fig. 4D).

3.7.5 Export the data

Upon completing the analysis of the initial sample, save the file as a template. This template will be named with the ".ast" suffix. Subsequently, apply this template consistently across all experimental conditions. See Note 18.

3.7.5.1 Visual inspection of Mask position

Conducting a thorough visual inspection of mask positioning on the images is crucial. This step is important to identify any potential errors resulting from software limitations. Apply the various masks to the bright field image or the specific fluorescence channel relevant to the "conjugates" population. Carefully select events where the mask alignment and recognition are precise. Manually annotate this curated population and save it, ensuring the accuracy of the data to be exported. See Note 19.

3.7.5.2 Data export and analysis

Once annotated, export the data to a statistical analysis software such as GraphPad Prism. In this software, create the necessary graphs and perform statistical tests to interpret the data. The presented data underwent an initial assessment to evaluate normal distribution using a Shapiro-Wilk test. Since data often deviate from normality, we opted for a non-parametric Mann-Whitney test to ascertain the statistical significance of our findings.

4. Concluding remarks

In this chapter, we aim to provide potential users with detailed and comprehensive methodologies for generating and analyzing cell-to-cell conjugates using IFC. We have developed a versatile analytical toolbox, structured with intuitive mask and feature combinations (Fig. 1), which enables researchers to explore diverse research questions. Our examples focus on the polarization of specific subcellular entities at both cell sides of the lytic IS established between NK cells and cancer cells. However, this toolbox is equally capable of analyzing various other cell components and different types of ISs.

A significant advantage of IFC over traditional forms of microscopy, such as confocal microscopy, lies in its capacity to handle larger sample sizes. Conventional microscopy typically restricts sample sizes to 20–50 cells, mainly due to time limitations in data acquisition. In contrast, IFC allows the analysis of at least 200 cell conjugates within a comparable, limited time frame. This substantial increase in sample size greatly enhances the robustness and statistical validity of the data. Remarkably, acquiring 5000 conjugates with IFC takes approximately only half an hour. However, it should be acknowledged that factors such as improper orientation or mask misalignment can render a considerable portion of the data unsuitable for analysis. This is particularly relevant in the context of analyzing the IS. Yet, the volume of data analyzed with IFC remains at least tenfold greater compared to what is achievable with classical fluorescence microscopy. Besides its high-throughput advantage, IFC also offers superior multiparametric capabilities. Whereas classical fluorescence microscopy often faces constraints due to color bleedthrough, IFC, as a cytometry-based method, circumvents these limitations by employing a compensation matrix. This significantly expands the number of available channels, thereby enhancing the capacity to conduct correlative analyses of multiple components of interest.

5. Notes

Note 1: As an alternative to genetic manipulation, specific cell-permeable dyes, such as SiR–actin (Emission at 674 nm, #sc001, tebubio) or SPY555-actin (Emission at 580 nm, #sc202, tebubio), can be utilized for F-actin labeling. The labeling protocol can be applied to either the target or the effector cell.

Note 2: Flow cytometry analysis, involving the labeling of CD3 and CD56 molecules, can be conducted on the day of the co-culture experiment to confirm the purity of primary NK cells ($CD3^{-}CD56^{+}$).

Note 3: We employed cells expressing a fluorescent actin reporter, which eliminates the need for additional staining to identify target cells. However, an alternative option is the use of fluorescent dyes for cytoplasmic staining. These dyes are available in various colors, ensuring compatibility with your fluorescent panel, options include the CellTracker™ Green CMFDA Dye (#C7025, Thermo Fisher Scientific), the CellTracker™ Orange CMRA (#C34551, Thermo Fisher Scientific) or CellTracker™ Deep Red Dye (#C34565, Thermo Fisher Scientific).

Note 4: Long incubation times with Trypsin-EDTA can cause cell damage and cleavage of membrane-bound proteins involved in conjugate formation (Lai et al., 2022; Tsuji et al., 2017). Thus, Trypsin-EDTA is used at low concentration for short time and after detachment, the target cells are placed in the incubator for at least 20 min for cell membrane recovery. Alternatively, other methods of cell detachment can be used, such as non-enzymatic solutions (e.g., Accutase). In our experimental conditions and with the cell line used in this protocol, we did not observe a difference in cells treated with Trypsin-EDTA or Accutase when working with HLA-A,B,C immuno-localization.

Note 5: CTV is a cell-permeable violet dye that reacts with amines, enabling visualization of the effector cell. Alternative dyes can also be employed and a few examples are provided in Note 3. In contrast, Zombie-NIR is an amine-reactive, non-permeable dye that penetrates cells with compromised membranes, thus exclusively labeling dead cells. NK cells are stained with Zombie-NIR to gate out conjugates formed with a dead NK in order to analyze only conjugates formed with healthy NK cells. The same staining protocol can be applied to the target cells.

Note 6: In this protocol, we used a cell line (MDA-MB-231 EmLA) constitutively expressing the fluorescent Emerald-LifeAct protein. Consequently, the mono-color staining for compensation of the HLA-A,B,C and CD63 channels were prepared using MDA-MB-231 WT cells.

Note 7: Conjugates are particularly delicate, especially prior to PFA fixation. Therefore, we avoid mixing the PFA directly with the cell pellet; instead, we add PFA cautiously and allow it to take effect. Even if the conjugates are stabilized post-PFA treatment, it is essential to handle them with care during resuspension and gentle mixing is still necessary to prevent dissociation of the conjugates (wide orifice pipette tips can be used).

Note 8: The extended depth of field (EDF™) is based on the Wavefront Coding™ technology developed by CDM Optics. This module combines specialized optics and image processing algorithms to achieve sharp focus off multiple cellular structures within a single plane. The depth of field is extended from 2.5 μm to 16 μm with the EDF mode. In this mode, the MTOC labeling was visible in approximately 60% of cells, whereas it was observed in fewer than 40% of cells without the EDF.

Note 9: We observed a potential alignment issue in EDF acquisition mode with the imaging flow cytometer, where slight rotations of cells between the two camera interception points may lead to less precise alignment. To address this, it is advisable to limit comparisons to channels within a single camera. When choosing the fluorescent panel where EDF mode is optimal for the structures being analyzed (e.g., spots or single object), it is crucial to consider this technical factor. As an example, we used actin in channel 2 and CD63 in channel 3 or MTOC in channel 4, ensuring these channels were on the same camera.

Note 10: We typically acquire 5×10^3 to 8×10^3 conjugates per sample. For some labeling, we aim to acquire a higher number of events due to the high number of un-analyzable conjugates. For example, for MTOC labeling, a substantial portion of cells may not exhibit labeling, possibly due to being out of focus, as a consequence of depth of field limitations. A pilot experiment to determine the optimal number of conjugates to acquire, ensuring accurate analysis, is recommended.

Note 11: While it is possible to utilize the matrix generated during acquisition, we highly recommend creating a new matrix via the Ideas wizard. The matrix table will be displayed. Validate the matrix; if red values are indicated, click on the value and refine the positive population to correct the matrix. Moreover, the mono-color files will be merged in a single file. Open the merged data and apply the matrix. Plot the intensity of each color against the others to assess compensation quality; if necessary, edit manually the matrix.

Note 12: A gate based on in-focus cells is not necessary in EDF acquisition mode, as all events are inherently focused.

Note 13: This paper details three illustrative examples conducted on the target cell side of the IS and one on the NK cell side. It is important to note that the analysis can be performed on both sides. Actin cytoskeleton labeling was employed to create a mask for surface molecules, cytosolic molecules, and spot-shaped structures on the target cell side.

Similarly, utilizing the cytoplasmic staining of CTV allows the same methodology to be applied on the NK cell side. For assessing single structures, CTV was used to mask the MTOC in NK cells; however, the identical approach can be utilized to detect the MTOC in the target cells by using actin cytoskeleton labeling to create the mask.

Note 14: The pixel size extension of the synaptic mask may need adjustment according to the labeling, the specific cell line area or the acquisition mode. It is crucial to visually assess this parameter across a substantial number of cells before finalizing a pixel number. In our study, we opted for an extension of 19 pixels, equivalent to one-third of the MDA-MB-231 cell. When EDF is employed for acquisition, the interface extension is set at 15 pixels to cover one-third of the MDA-MB-231 cells, as masks are smaller in EDF mode.

Note 15: If a broader coverage of the plasma membrane is required for another application, one can utilize the "Dilate(M02,2)" operation instead of "Dilate(M02,0)", this will extend the dilatation by 2 pixels. However, it is crucial to extend the dilation to the interface mask as well, achieved by applying "Dilate(Interface(M02, M06, 19),2)."

Note 16: We employed the actin channel for surface molecule mask generation. In instances where cytosolic labeling is absent, pre-labeling with a cytosolic dye, such as CMFDA (#C2925, Thermo Fisher Scientific) or alternatively, using the surface molecule channel to create the mask is possible. However, in our specific experiments, the mask based on the HLA labeling, extended beyond the target area due to a weak HLA-A,B,C signal in the NK cells. Additionally, we observed that the Dilate function may not consistently cover the surface molecule labeling effectively. An alternative is the object function, although the resulting mask is smaller than Dilate. Consequently, it may be necessary to first create the mask object "Object(M04,Ch04,Tight)" and subsequently dilate the object mask using "Dilate(Mask Object,2)" to dilate the object by 2 pixels and properly cover the surface molecule labeling.

Note 17: When employing EDF for acquisition, the interface extension is set at 15 pixels to cover one-third of the target cell, as masks are smaller in EDF mode. In non-EDF mode, the width index is set at 19 pixels.

Note 18: The creation of the data analysis file (daf) from the raw image can be time consuming. An alternative, once the compensation matrix and analysis template are generated, is to use the batch process tool to create all the corresponding .daf files from one experiment before the analysis.

Note 19: To facilitate result visualization, the features can be emphasized within the image. In the image gallery properties, navigate the "views" tab, and activate the "display feature value on image" option to incorporate the feature values onto the picture.

Acknowledgments

C.T.'s lab is generously supported by La Fondation Cancer (Luxembourg, ACTIMMUNE, FC/2019/02) and Think Pink Lux. A.M.B. is recipient of a PhD fellowship provided by the Fonds De La Recherche Scientifique (FNRS; ACTIPHAGY; Televie grant #34979532). We wish to express our gratitude to the National Cytometry Platform (NCP), and particularly Maira Konstantinou, for her assistance in the generation and analysis of IFC data. The NCP gratefully acknowledge its supports from Luxembourg's Ministry of Higher Education and Research (MESR).

Bibliography

Ahmed, F., et al. (2009). Numbers matter: Quantitative and dynamic analysis of the formation of an immunological synapse using imaging flow cytometry. *Journal of Immunological Methods, 347*(1–2), 79–86.

Al Absi, A., et al. (2018). Actin cytoskeleton remodeling drives breast cancer cell escape from natural killer-mediated cytotoxicity. *Cancer Research, 78*(19), 5631–5643.

Biolato, A. M., et al. (2020). Actin remodeling and vesicular trafficking at the tumor cell side of the immunological synapse direct evasion from cytotoxic lymphocytes. *International Review of Cell and Molecular Biology, 356*, 99–130.

Cantoni, C., et al. (2020). Escape of tumor cells from the NK cell cytotoxic activity. *Journal of Leukocyte Biology, 108*(4), 1339–1360.

Dominical, V., Samsel, L., & McCoy, J. P., Jr. (2017). Masks in imaging flow cytometry. *Methods, 112*, 9–17.

Filali, L., et al. (2022). Ultrarapid lytic granule release from CTLs activates Ca(2+)-dependent synaptic resistance pathways in melanoma cells. *Science Advances, 8*(7), eabk3234.

Khazen, R., et al. (2016). Melanoma cell lysosome secretory burst neutralizes the CTL-mediated cytotoxicity at the lytic synapse. *Nature Communications, 7*, 10823.

Lai, T. Y., et al. (2022). Different methods of detaching adherent cells and their effects on the cell surface expression of Fas receptor and Fas ligand. *Scientific Reports, 12*(1), 5713.

Mace, E. M., et al. (2014). Cell biological steps and checkpoints in accessing NK cell cytotoxicity. *Immunology and Cell Biology, 92*(3), 245–255.

Markey, K. A., & Gartlan, K. H. (2019). Imaging flow cytometry to assess antigen-presenting-cell function. *Current Protocols in Immunology, 125*(1), e72.

McKenzie, B., & Valitutti, S. (2023). Resisting T cell attack: Tumor-cell-intrinsic defense and reparation mechanisms. *Trends in Cancer, 9*(3), 198–211.

Ockfen, E., et al. (2023). Actin cytoskeleton remodeling at the cancer cell side of the immunological synapse: Good, bad, or both? *Frontiers in Immunology, 14*, 1276602.

Quatrini, L., et al. (2021). Human NK cells, their receptors and function. *European Journal of Immunology, 51*(7), 1566–1579.

Riedl, J., et al. (2008). Lifeact: A versatile marker to visualize F-actin. *Nature Methods, 5*(7), 605–607.

Sivori, S., et al. (2020). Inhibitory receptors and checkpoints in human NK cells, implications for the immunotherapy of Cancer. *Frontiers in Immunology, 11*, 2156.

Tsuji, K., et al. (2017). Effects of different cell-detaching methods on the viability and cell surface antigen expression of synovial mesenchymal stem cells. *Cell Transplantation*, *26*(6), 1089–1102.

Viswanath, D. I., et al. (2017). Quantification of natural killer cell polarization and visualization of synaptic granule externalization by imaging flow cytometry. *Clinical Immunology*, *177*, 70–75.

Wabnitz, G. H., et al. (2015). InFlow microscopy of human leukocytes: A tool for quantitative analysis of actin rearrangements in the immune synapse. *Journal of Immunological Methods*, *423*, 29–39.

Wurzer, H., et al. (2021). Intrinsic resistance of chronic lymphocytic leukemia cells to NK cell-mediated lysis can be overcome in vitro by pharmacological inhibition of Cdc42-induced actin cytoskeleton remodeling. *Frontiers in Immunology*, *12*, 619069.

Zuba-Surma, E. K., et al. (2007). The ImageStream system: A key step to a new era in imaging. *Folia Histochemica et Cytobiologica*, *45*(4), 279–290.

CHAPTER FIVE

Quantifying force-mediated antigen extraction in the B cell immune synapse using DNA-based tension sensors

Hannah C.W. McArthur[a,†], Anna T. Bajur[a,b,†], and Katelyn M. Spillane[a,b,c,*]

[a]Department of Physics, King's College London, London, United Kingdom
[b]Randall Centre for Cell & Molecular Biophysics, King's College London, London, United Kingdom
[c]Department of Life Sciences, Imperial College London, London, United Kingdom
[*]Corresponding author: e-mail address: k.spillane@imperial.ac.uk

Contents

1. Introduction	100
2. Materials	102
2.1 DNA oligomer sequences	102
2.2 Preparation of antigen-functionalized DNA tension sensor	105
2.3 Imaging chamber preparation	105
2.4 Planar lipid bilayer formation and calibration	106
2.5 Preparing and adding primary naive B cells to planar lipid bilayers	106
2.6 Staining B cells for imaging	107
2.7 Equipment and software	107
3. Methods	108
3.1 Preparing antigen-functionalized DNA tension sensors	108
3.2 Cleaning glass coverslips	109
3.3 Assembling imaging chambers	111
3.4 Planar lipid bilayer formation	112
3.5 Lipid calibration standards	115
3.6 Measuring antigen extraction by primary B cells at fixed timepoints	118
3.7 Quantitative image analysis	120
4. Notes	121
5. Conclusions	124
Acknowledgments	124
References	124

[†] Equal contribution.

Methods in Cell Biology, Volume 193
ISSN 0091-679X
https://doi.org/10.1016/bs.mcb.2024.03.002

Copyright © 2025 Elsevier Inc.
All rights are reserved, including those
for text and data mining, AI training,
and similar technologies.

Abstract

B cells exert pulling forces against antigen-presenting cells (APCs) to extract antigens for internalization. The application of tugging forces on B cell receptor (BCR)-antigen bonds promotes discrimination of antigen affinities and sensing of APC physical properties. Here, we describe a protocol for preparing antigen-functionalized DNA tension sensors for quantifying force-mediated antigen extraction in the B cell immune synapse. We describe how to attach the sensors to planar lipid bilayers, quantify their surface density, use them to stimulate B cell activation, and analyze the efficiency of antigen extraction in fixed cells by fluorescence microscopy and image analysis. These techniques should be broadly applicable to studies of force-mediated transfer of molecules in cell-cell contacts.

1. Introduction

The production of high-affinity antibodies during an immune response begins with direct contacts between B cells and antigen-presenting cells (APCs). B cells scan APC surfaces for antigens using actin-based membrane protrusions that are densely packed with B cell receptors (BCRs) (Jung, Wen, Altman, & Ley, 2021; Saltukoglu et al., 2023). Specific binding interactions between BCRs and antigens trigger intracellular signaling cascades that result in the formation of a structured cell-cell interface, called an immune synapse, between the B cell and APC (Batista, Iber, & Neuberger, 2001). In the synapse, B cells apply active tugging forces on BCR-antigen bonds by contracting the actomyosin cytoskeleton (Natkanski et al., 2013). These forces allow B cells to extract antigens from the APC surface and internalize them into compartments for processing (Batista & Neuberger, 2000). B cells present antigenic peptides with major histocompatibility complex (MHC) II molecules to facilitate interactions with T cells, which provide contact-dependent signals that promote B cell survival, proliferation, and plasma cell differentiation (Crotty, 2015). These events also drive the evolution of high-affinity BCRs through mutation and selection by T cells in germinal centers (Victora & Nussenzweig, 2022), where B cell clones compete for T cell help based on the amount of antigen they internalize and present (Schwickert et al., 2011; Suzuki, Grigorova, Phan, Kelly, & Cyster, 2009). The active tugging forces that B cells exert against APCs to capture antigens thus contribute to the affinity maturation of antibodies (Jiang & Wang, 2023).

Antigens are typically presented on APC membranes in the form of immune complexes, which contain multiple antigens that are bound by antibodies and opsonized with complement. Immune complexes can be captured on APC membranes through one of several antigen receptors including Fcγ receptors (FcγRs) that bind the constant region of IgG antibodies (Iliopoulou et al., 2024) and complement receptors (CRs) that bind complement fragments (Heesters et al., 2013). When B cells pull on antigens in the immune synapse, the forces are propagated through a series of non-covalent interactions that connect the B cell and APC membranes including bonds between BCRs and antigens (tugging complexes), and bonds between antigens and antigen receptors (tethering complexes) (Spillane & Tolar, 2018a). BCRs bind monovalent antigens over a wide range of affinities ($K_d \approx 1\,\mu M$ to $100\,pM$) (Batista & Neuberger, 1998, 2000), which is extended to hundreds of μM for multivalent interactions (Schmidt et al., 2013; Wedemayer, Patten, Wang, Schultz, & Stevens, 1997). The antigen receptors on APCs bind antigens with a narrower range of K_d values, spanning 0.5 to $5\,\mu M$ for FcγR–IgG interactions (Bruhns et al., 2009) and 0.1 to $1\,\mu M$ for CR–complement interactions (Bajic, Yatime, Sim, Vorup-Jensen, & Andersen, 2013; Dustin, 2016; Sarrias et al., 2001). B cell discrimination of antigen affinities in the immune synapse therefore occurs through competitive rupture of tugging and tethering complexes, the lifetime of which is further influenced by the stiffness of B cell and APC membranes (Spillane & Tolar, 2017, 2018a).

Advances in planar lipid bilayer, multidimensional imaging, and molecular force probe technologies have made it possible to quantify antigen internalization in the B cell synapse and investigate how the relative stabilities of tugging (BCR–antigen) and tethering (antigen–APC) bonds influence the sensitivity and specificity of antigen capture. Here, we provide a detailed protocol to conduct such experiments. In these experiments, a DNA duplex tethers antigen molecules to the bilayer. The stability of BCR–antigen bonds can be tuned by attaching antigens that bind the BCR with different affinity. The stability of DNA duplexes mimicking antigen–APC bonds can be tuned by changing the DNA sequence and geometry (Wang & Ha, 2013). We further provide open-access ImageJ macros to perform automated and semi-automated image processing and analysis. These are available on our GitHub page (https://github.com/SpillaneLab). A summary of the entire workflow and methods to be described is given in Fig. 1.

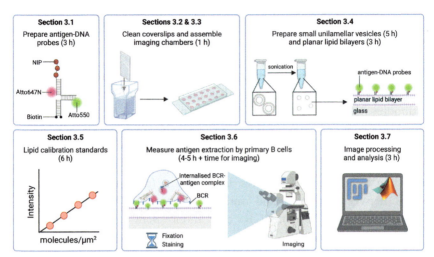

Fig. 1 Scheme summarizing the workflow detailed in the text for measuring B cell antigen internalization from planar lipid bilayers using DNA probes. We describe the preparation of antigen-DNA force probes (Section 3.1), glass-bottomed imaging chambers (Sections 3.2 and 3.3), and functionalized planar lipid bilayers (Section 3.4). We also explain how to determine antigen-DNA surface densities on bilayer substrates (Section 3.5), and image (Section 3.6) and quantify antigen internalization by B cells using fluorescence microscopy (Section 3.7). Approximate timescales for each step are provided.

2. Materials

This protocol describes the assembly of DNA tension sensors and their conjugation to three NIP or NP haptens, which bind specifically to the B1-8i BCR with different affinities ($K_{d,NIP} = 0.33\,\mu M$ and $K_{d,NP} = 2.2\,\mu M$; Natkanski et al., 2013). The presentation of three happens per sensor promotes multivalent interactions with the BCR, and the valency can be easily tuned by changing the number of Uni-Link modifiers per DNA construct. For conjugating protein antigens to DNA, the reader is directed to Spillane and Tolar (2018b). Alternatively, other conjugation methods including SNAP (Keppler et al., 2003) and Halo (Los et al., 2008) fusion tags can be used to attach proteins to O^6-benzylguanine- and HaloTag ligand-modified DNA sensors, respectively.

2.1 DNA oligomer sequences

Each sensor comprises four single-stranded DNA oligomers. Upon assembly, the sensors have two 24-bp handles that are linked by a thermodynamically weaker 20-bp duplex. The upper handle is modified with amino

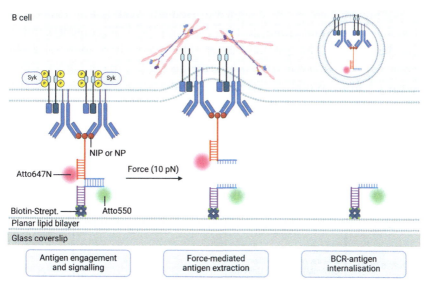

Fig. 2 Schematic representation of DNA-based force sensors. The ladders represent DNA. When the B cell applies a force exceeding the DNA rupture threshold, the central DNA duplex unzips, allowing the B cell to internalize the BCR-antigen-DNA-Atto647N complex. Color coding of DNA oligos—red: 24-bp upper handle; purple: 24-bp lower handle; blue: 20-bp "tethering bond." The corresponding sequences are provided in Sections 2.1.1 and 2.1.2.

groups for covalent attachment of the DNA to NHS-coupled haptens (NIP or NP; antigen) and an Atto647N fluorophore for monitoring antigen uptake by B cells. The lower handle contains two biotin modifications to establish strong binding between the DNA and streptavidin-functionalized bilayers (refer to Fig. 2 and compare the sequences in Sections 2.1.1 and 2.1.2). The lower strand of the 20-bp duplex has an Atto550 fluorophore for monitoring total antigen bound by the B cell.

Forces exerted on BCR-antigen bonds transmit across the DNA to the bilayer surface. In the direction of the pulling force, the 20-bp duplex has an unzipping configuration, making it prone to melt abruptly at forces ranging from 10 to 20 pN, depending upon the % GC content (Albrecht et al., 2003) (*see Note 1*). Rupture of the 20-bp duplex releases antigen from the surface for internalization. During this process, the two 24-bp handles remain intact because they are either unaffected by the force (upper handle) or have a more robust shear geometry (Albrecht et al., 2003) that is further stabilized by two biotin-streptavidin bonds (lower handle) (Stayton et al., 1999). Consequently, the DNA sensor operates as a competitive binding assay between the BCR-antigen "tugging bond" and the 20-bp dsDNA "tethering bond" (Fig. 2).

We present two sensors designed to establish weak (20-bp tether with 20% GC content) and strong (20-bp tether with 100% GC content) connections between the antigen and bilayer. The sensors have identical 24-bp handle sequences but different 20-bp tether sequences. In our nomenclature, borrowed from Woodside et al. (2006), the sequence 20R100 UH24 signifies two domains: (20R100) a 20-bp quasi-random (R) sequence with 100% GC content and (UH24) a 24-bp upper handle. The notation "C" denotes the reverse complement (e.g., 20R100C is the reverse complement of 20R100). LH24 represents the 24-bp lower handle sequence.

2.1.1 Weak DNA tether (20% GC content)

Oligo	5' modification	Sequence	3' modification
1. 20R20 UH24	Uni-Link Amino Modifier (UniAmM)-UniAmM-UniAmM	tca cga cag gtt cct tcg cat cga tat tct atg tta cta tgt aa	
2. LH24 20R20C	Atto550	tta cat agt aac ata gaa taa cat gtg tga cga aca ctc att ta	Biotin
3. UH24C	Atto647N	tcg atg cga agg aac ctg tcg tga	
4. LH24C	Biotin	taa atg agt gtt cgt cac aca tgt	

Blue: 20-bp tether sequence; red: 24-bp upper handle; purple: 24-bp lower handle.

2.1.2 Strong DNA tether (100% GC content)

Oligo	5' modification	Sequence	3' modification
1. 20R100 UH24	UniAmM-UniAmM-UniAmM	tca cga cag gtt cct tcg cat cga cgc cgc ggg ccg gcg cgc gg	
2. LH24 20R100C	Atto550	ccg cgc gcc ggc ccg cgg cga cat gtg tga cga aca ctc att ta	Biotin
3. UH24C	Atto647N	tcg atg cga agg aac ctg tcg tga	
4. LH24C	Biotin	taa atg agt gtt cgt cac aca tgt	

Blue: 20-bp tether sequence; red: 24-bp upper handle; purple: 24-bp lower handle.

2.2 Preparation of antigen-functionalized DNA tension sensor

- Single-stranded DNA oligomers purified by HPLC, resuspended to 100 μM in duplex buffer (store at $-20\,^\circ$C)
- Nuclease-free duplex buffer: 30 mM HEPES, pH 7.5, 100 mM potassium acetate (Integrated DNA Technologies; store at $4\,^\circ$C)
- Magnesium chloride ($MgCl_2$; 100 mM in double-distilled water (ddH_2O))
- Zeba spin desalting columns, 7 kDa MWCO, 0.5 mL (Thermo, 89882)
- 1.5 mL DNA LoBind tubes (Eppendorf, 0030108051)
- 4-Hydroxy-3-iodo-5-nitrophenylacetic active ester (NIP-Osu; Biosearch Technologies, N-1080-100)
- 4-Hydroxy-3-nitrophenylacetic acid active ester (NP-Osu; Biosearch Technologies, N-1010-100)
- Anhydrous DMSO (Invitrogen, D12345)
- Labeling buffer: 0.1 M $NaHCO_3$, pH 8.3

2.3 Imaging chamber preparation

- 95% sulfuric acid (Fisher Scientific, 10391313)
- 30% hydrogen peroxide (VWR, 23615.248)
- Absolute ethanol (Merck, 32221)
- 24 × 50 mm cover glass No. 1.5 (Fisher Scientific, 11836933)
- Sodium bicarbonate (Fisher Scientific, BP328-500)
- Hellmanex® III solution (Hellma UK Ltd.; sourced from Merck, Z805939-1EA)
- 99.5 + % isopropanol (IPA) (Merck, 190764)
- Potassium hydroxide pellets (Fisher Scientific, 10388103)
- KimWipes (Kimberley-Clark)
- 8-well Lab-Tek chambered coverglass (Thermo Fisher Scientific, 155411) or 1-well Lab-Tek chambered coverglass (Thermo Fisher Scientific, 155361), with coverslip and adhesive removed (*see Note 2*)
- Clear adhesive tape (Sellotape, Scotch Tape, or similar)
- Polystyrene weighing boats
- 1 mL plastic syringes
- CultureWell™ 3–10 μL re-useable silicone gaskets (Grace Bio-Labs, 103250) (*see Note 3*)
- Dow SYLGARD™ 170 fast-cure silicone (Ellsworth Adhesives, PF170FC200S)

- 4 mL glass threaded vials with attached caps (Fisher Scientific, 14-955-334, or similar)
- 18G × 40 mm blunt fill needles (BD, 303129)

2.4 Planar lipid bilayer formation and calibration

- 18:1 (Δ9-*cis*) PC DOPC (Avanti, 850375C-200 mg; 25 mg/mL chloroform solution) (store at $-20\,^\circ$C)
- 18:1 Biotinyl Cap DOPE (biotin-DOPE; Avanti, 870273C-25 mg; 10 mg/mL chloroform solution) (store at $-20\,^\circ$C)
- 18:1 Liss Rhod PE (Avanti, 810150C-1 mg; 1 mg/mL chloroform solution) (store at $-20\,^\circ$C)
- Chloroform (Merck, 288306)
- Absolute ethanol (Merck, 32221)
- Phosphate buffered saline (PBS) (Invitrogen, AM9625)
- Bemis Parafilm M Laboratory Wrapping Film (Fisher Scientific, 11772644)
- Bovine serum albumin (BSA; Fisher Scientific, PB1600-100), 600 μg/mL in PBS
- Streptavidin (Merck, S4762-1 mg), 6 μg/mL in PBS
- Biotinylated DNA-antigen probes (see Section 3.1)

2.5 Preparing and adding primary naive B cells to planar lipid bilayers

1. B6.129P2(C)-*Igh*^{tm2Cgn}/J mice (B1-8i mice; The Jackson Laboratory)
2. MACS buffer: PBS, pH 7.4, 2 mM EDTA, 0.5% (w/v) BSA (store at 4 $^\circ$C)
3. 2 mL disposable plastic syringe without needle (BD Plastipak or similar)
4. 6-well multiple well plates, sterile (Fisher Scientific, 10578911; or similar)
5. 70 μm sterile cell strainers (Fisher Scientific, 11517532; or similar)
6. eBioscience 1× RBC lysis buffer (Thermo Fisher Scientific, 00-4333-57)
7. CD43 (Ly-48) MicroBeads, mouse (Miltenyi Biotec, 130-049-801)
8. MidiMACS Separator (Miltenyi Biotec, 130-042-302) with MACS MultiStand (Miltenyi Biotec, 130-042-303)
9. LD columns (Miltenyi Biotech, 130-042-901)
10. Trypan Blue solution, 0.4% (Thermo Fisher Scientific, 15250061)
11. 15 mL Falcon conical centrifuge tubes, polypropylene (Fisher Scientific, 10773501; or similar)

12. Full RPMI: Roswell Park Memorial Institute (RPMI) medium 1640, no glutamine (Thermo Fisher Scientific, 31870025) supplemented with 10% fetal bovine serum (FBS) (Thermo Fisher Scientific, 10500064), 1% MEM non-essential amino acids solution (Thermo Fisher Scientific, 11140050), 2 mM L-glutamine (Thermo Fisher Scientific, 25030149), 50 μM 2-mercaptoethanol (Thermo Fisher Scientific, 21985023), 100 U/mL penicillin and 100 μg/mL streptomycin (Thermo Fisher Scientific, 15140148).
13. Imaging buffer: Hank's Balanced Salt Solution (HBSS with Ca^{2+} and Mg^{2+}; Thermo Fisher Scientific, 14025092) plus 0.1% BSA (store at 4 °C)

2.6 Staining B cells for imaging

1. Paraformaldehyde 16% *w/v* aqueous solution, methanol free (Thermo Fisher Scientific, 043368.9M), aliquoted and stored at −20 °C
2. Normal mouse serum (Jackson ImmunoResearch, 015-000-120), aliquoted and stored at −20 °C
3. Brilliant Violet 421 (BV421) anti-mouse/human CD45R/B220 antibody (BioLegend, 103251)

2.7 Equipment and software

1. Vertical glass staining jar, ×2 (Fisher Scientific, 15293867)
2. 2 L glass beaker (Fisher Scientific, 15459083)
3. 100 mL crystalizing dish without spout (VWR, KAVAN632412624170)
4. Argon or nitrogen
5. Coverslip mini-rack, for 8 coverslips (Thermo Fisher Scientific, C14784)
6. 500 μL Gastight syringe model 1750, large removable needle, 22 gauge, 2 in., point style 3 (Hamilton, 81265) (for DOPC lipids)
7. 100 μL Gastight syringe model 1710, small removable needle, 22 gauge, 2 in., point style 3 (Hamilton, 81065) (for biotin-DOPE lipids)
8. 10 μL Gastight syringe model 1701, small removable needle, 32 gauge, 2 in., point style 3 (Hamilton, 80014) (for Liss Rhodamine PE lipids)
9. CryoTubes (Fisher Scientific, 10674511)
10. 250 mL glass squat-form beaker (Fisher Scientific, 15409083; or similar)
11. Refrigerated bench-top centrifuge (Eppendorf 5424R or similar)
12. Bath sonicator (Fisherbrand FB 15050 or similar)
13. Vortex mixer (Stuart Scientific SA8 or similar)

14. Variable speed thermal mixer (Eppendorf ThermoMixer C or similar)
15. Analytical balance (Sartorius Entris II or similar)
16. Oven
17. Diaphragm vacuum pump (Vacuubrand ME 1 or similar)
18. Desiccator (Nalgene transparent polycarbonate classic design desiccator, or similar)
19. Portable aspirator (Integra Vacusafe or similar)
20. Bright-Line Hemacytometer (Merck Z359629-1EA, or similar)
21. Microvolume spectrophotometer (Thermo Fisher Scientific NanoDrop or similar)
22. Thermal cycler (ProFlex PCR system or similar)
23. Optical power meter (Thorlabs PM100D) with photodiode power sensor (Thorlabs S130C)
24. Inverted epifluorescence microscope with a 100× objective, motorized xyz-movement, and autofocusing (Nikon TiE or similar)
25. Image acquisition software capable of z-stack imaging such as MicroManager (Edelstein, Amodaj, Hoover, Vale, & Stuurman, 2010)
26. ImageJ/Fiji (Schindelin et al., 2012)
27. Optional: UV/ozone plasma cleaner (Plasma System Femto, Diener Electronic, or similar)

3. Methods

3.1 Preparing antigen-functionalized DNA tension sensors

The sensors can be assembled, conjugated to haptens, purified, and stored in 1 day.

1. In a PCR tube, mix 7.5 μL of each 100 μM single-stranded DNA oligomer and 0.6 μL of 100 mM $MgCl_2$ (final $MgCl_2$ concentration of 2 mM) (*see Note 4*). Anneal the mixed oligonucleotides by heating to 94 °C for 2 min in a thermocycler and transferring to the bench top to gradually cool to room temperature over 30 min, protecting them from light. The concentration of annealed DNA sensor will be 24.5 μM.
2. Equilibrate one Zeba spin desalting column with labeling buffer following the manufacturer's instructions.
3. Exchange the annealed DNA sensor into labeling buffer by spinning it through the desalting column.
4. Weigh a small amount of NIP-Osu or NP-Osu hapten into a 1.5 mL microcentrifuge tube using an analytical balance, and dissolve in anhydrous DMSO to make a 10 mM solution. Add 4.5 μL of the 10 mM

hapten to the DNA sensor solution (a 60-fold molar excess). Incubate for 2 h at room temperature, with mixing (550 rpm), using a thermal mixer.

5. Equilibrate two Zeba spin desalting columns with PBS following the manufacturer's instructions.

6. Spin the sensor sequentially through the two columns to remove unbound hapten.

7. Measure the sensor concentration and labeling ratio using a Nanodrop. Use the Atto550 absorbance at 550 nm ($\varepsilon_{max} = 120,000\,M^{-1}\,cm^{-1}$) to determine the DNA sensor molar concentration and the absorbance at 430 nm to determine the NIP or NP molar concentration. The extinction coefficients at 430 nm are $5000\,M^{-1}\,cm^{-1}$ for NIP and $4230\,M^{-1}\,cm^{-1}$ for NP. The expected labeling ratio is 3:1 NIP: DNA. Typical DNA sensor concentrations following desalting are $10–15\,\mu M$.

8. Freeze NIP_3- and NP_3-DNA probes, at $10–15\,\mu M$ concentration, in single-use $2\,\mu L$ aliquots at $-20\,°C$ (*see Note 5*).

3.2 Cleaning glass coverslips

We prepare bilayers by vesicle fusion on glass coverslips. The vesicles fuse to form a laterally fluid bilayer upon contact with a hydrophilic glass surface. The cleanliness of the glass substrate is crucial to prepare reproducible bilayers with constant diffusive properties. For optimum results, glass should be cleaned immediately before preparing the bilayers. We describe three different methods to clean glass coverslips that produce similarly good results.

3.2.1 Bath sonication and plasma cleaning

This method is the easiest and safest, and is recommended if a UV/ozone plasma cleaner is available. This protocol is adapted from Andrecka, Spillane, Ortega-Arroyo, and Kukura (2013).

- Load glass coverslips in a mini-rack and submerge in a 250 mL squat-form glass beaker filled with 50:50 ddH_2O:isopropanol. Bath sonicate for 10 min.
- Dispose of the ddH_2O:isopropanol solution, submerge the mini-rack with coverslips in ddH_2O, and bath sonicate for 10 min.
- Dispose of the ddH_2O, submerge the mini-rack with coverslips in 1 M potassium hydroxide, and bath sonicate for 10 min.
- Dry the coverslips under a stream of inert gas.
- Expose the coverslips to UV/ozone for 8 min at 50 W power using a plasma cleaner.

3.2.2 Piranha cleaning of glass coverslips

Piranha solution is extremely corrosive to organic material and should be used only to remove trace amounts of organic residues from substrates. Ensure risk assessments are in place prior to use, users are trained and wear full PPE (lab coat, gloves, safety glasses), and work is performed in a chemical fume hood. Prepare only as much piranha solution as needed. Never store piranha solution in a capped bottle due to the risk of gas overpressure and explosion. It should be disposed of after use following local safety guidelines.

1. Using a permanent marker, label 30 and 45 mL volume levels on the outside of the vertical glass staining jar.
2. Place the staining jar into the crystallization dish (in case of spillover).
3. Using the marker positions as a guide, slowly pour 30 mL of sulfuric acid followed by 15 mL of 30% hydrogen peroxide into the staining jar.
4. Place glass coverslips into the piranha solution individually using tweezers, ensuring the coverslips do not touch each other. Leave coverslips in the solution for 12 min.
5. Using tweezers, transfer the coverslips into a second staining jar containing ddH$_2$O.
6. Dispose of piranha solution following the local safety guidelines.
7. Wash coverslips 10× with ddH$_2$O, followed 3× with absolute ethanol. Store coverslips in ethanol until ready to use.
8. Dry coverslips under a stream of inert gas.

3.2.3 Hellmanex III cleaning of glass coverslips

Hellmanex III is a less rigorous but safer alternative to piranha solution. It is essential for making good bilayers that all Hellmanex III is washed from the coverslips. Hellmanex III is a basic solution (pH 11.6 at 0.5%), so users should wear appropriate PPE when handling and neutralize the solution before disposal.

1. Load glass coverslips in a mini-rack and submerge them in a 250 mL squat-form glass beaker filled with 0.5% (v/v) aqueous solution of Hellmanex III. Bath sonicate for 20 min.
2. Rinse the coverslips 10× with ddH$_2$O.
3. Bath sonicate the coverslips in ddH$_2$O for 15 min.
4. Repeat steps 2 and 3, using fresh ddH$_2$O for the washing and sonication steps.
5. Submerge the mini-rack with coverslips in isopropanol and bath sonicate for 15 min.
6. Rinse the coverslips 3× with isopropanol.

Quantifying force-mediated antigen extraction in the B cell immune synapse

7. Rinse the coverslips 3× with absolute ethanol. Store the coverslips in absolute ethanol until ready to use.
8. Dry coverslips under a stream of inert gas.

3.3 Assembling imaging chambers

We present protocols to prepare imaging chambers with either 250 or 10 µL wells. The larger volume is easier to work with while the smaller volume is useful when DNA tension sensors or cell numbers are limiting. The procedure is shown in Movie S1 in the online version at https://doi.org/10.1016/bs.mcb.2024.03.002.

1. Use tweezers to remove the glass coverslip and adhesive from the 1-well (for 10 µL volume) or 8-well (for 250 µL volume) Lab-Tek chamber. Ensure that no glass fragments or adhesive remains.
2. If using the 1-well Lab-Tek chamber, cut a strip of CultureWell 3–10 µL re-usable silicone gaskets, place onto an ethanol-cleaned bench, and use a piece of clear adhesive tape to remove lint from both sides. Invert a clean glass coverslip onto the gaskets and press gently with tweezers to form a good seal (Fig. 3, 1 and 2).
3. Mix equal volumes of the black and white components of the fast-cure silicone into a weighing boat (Fig. 3, 3 and 4)
4. Mix the fast-cure silicone well to form a uniform gray mixture (Fig. 3, 5).
5. Suck the fast-cure silicone into a 1 mL syringe, attach a blunt fill needle, and dispense the silicone around the 1- or 8-well chamber edge. Use excess silicone for a good seal (Fig. 3, 6–9).
6. Press the coverslip onto the chamber gently with tweezers to ensure the coverslip is flat. For 1-well Lab-Tek chambers, use a coverslip with the CultureWell gaskets already adhered, and be sure that the gaskets face into the Lab-Tek chamber (Fig. 3, 10).
7. Remove excess silicone using a KimWipe.
8. Allow the silicone to cure at room temperature for at least 20 min. Use the remaining silicone glue in the weighing boat as an indicator of when the silicone is fully cured.
9. It is likely that some silicone will invade the interior walls of the chambers. This will not impact the health or behavior of the cells. However, if desired, this silicone can be removed with tweezers once it is cured. Avoid contacting the glass coverslip with the tweezers, which can scratch the coverslip surface.
10. If needed, use a KimWipe and absolute ethanol to clean the bottom surface of the coverslip.

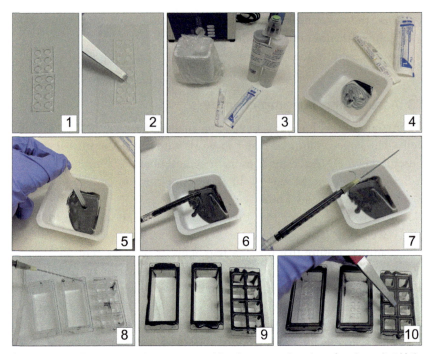

Fig. 3 Images showing step-by-step assembly of imaging chambers for planar lipid bilayers. (1–2) Cut a strip of silicone gaskets onto a bench and attach a cleaned glass coverslip. (3–5) Next, pour some Dow Corning 170 fast-cure silicone elastomer into a weighing boat, and mix thoroughly until the black and white components have come together to form a uniform gray color. (6–7) Take the glue up into the syringe and attach the needle. (8–9) Trace around the chamber edges with the glue, making sure to use enough glue such that the edges are nearly overflowed. (10) Place a clean glass coverslip onto the glue, pushing down gently with tweezers over the glued regions to form a tight seal. Remove excess glue from the sides gently with a Kim wipe. Keep the weighing boat with remaining glue to determine when the glue has dried fully, at which point SUVs can be added to the wells.

3.4 Planar lipid bilayer formation

We present protocols for the preparation of small unilamellar vesicles (SUVs) and their fusion onto glass coverslips. This procedure yields 2 mL of SUVs with a lipid concentration of roughly 4 mg/mL. The composition of SUVs consists of 99 mol% DOPC and 1 mol% biotinyl-cap DOPE, the latter being essential for the attachment of biotinylated DNA sensors through streptavidin. This composition enables the attachment of physiological densities of antigens/DNA sensors on the bilayer surface, ranging from tens to hundreds of molecules per μm^2, by titrating the concentration of antigen/DNA sensor.

Maintaining a clean environment is essential for the success of this protocol. Use freshly filtered buffers, and cover samples to prevent the accumulation of dust particles. The protocol is based on previously published work (Nair, Salaita, Petit, & Groves, 2011).

3.4.1 SUV preparation

1. Warm lipid chloroform stock solutions to room temperature (at least 30 min).
2. Clean a 4 mL glass threaded vial by rinsing first with absolute ethanol and then with chloroform. Dispose of solvents through appropriate routes.
3. Clean Hamilton syringes 10× with absolute ethanol, followed 10× with chloroform.
4. Use a 500 μL Hamilton syringe to transfer 315.5 μL of 25 mg/mL DOPC and a 100 μL Hamilton syringe to transfer 11.2 μL of 10 mg/mL biotin-DOPE lipids to the clean glass vial (*see Note 6*).
5. Dry lipids to a film using a stream of inert gas (argon or nitrogen). Remove residual chloroform by placing the vial into a desiccator connected to a vacuum pump, and placing under vacuum for at least 3 h.
6. While the lipids are drying, filter 5 mL PBS through a 0.22 μm sterile syringe filter into a 15-mL Falcon tube, place the tube into the vacuum pump-connected desiccator, and de-gas under vacuum until bubbling or "boiling" no longer occurs (at least 30 min).
7. Add 2 mL of de-gassed PBS to the lipid film, flush with inert gas to displace oxygen, cap the vial, and seal with parafilm.
8. Vortex the lipid mixture for several minutes until the film disappears, yielding a milky solution.
9. Bath sonicate the lipid mixture in 30-min intervals (usually 1 h in total), with a 1-min stop time between each sonication cycle, until the mixture becomes transparent and slightly blue in color, indicating a small and uniform vesicle size (*see Note 7*).
10. Keep 15 μL of the pre-spin SUV mixture for later UV–vis analysis.
11. Transfer the SUVs to a sterile 1.5 mL microcentrifuge tube and spin for 10 min at $16,000 \times g$ at 4 °C.
12. Transfer the supernatant containing SUVs to a CryoTube and flush with inert gas to displace oxygen. Seal the tube with parafilm.

13. Measure the absorption, A, of the pre- and post-spin SUV samples at 230 nm using a NanoDrop and calculate the final lipid concentration, C, using Eq. (1):

$$C = \frac{A_{\text{post-spin}}}{A_{\text{pre-spin}}} \times 4 \text{ mg/mL} \qquad (1)$$

14. Store the SUVs under inert gas at 4 °C for up to 1 month.

3.4.2 Planar lipid bilayer preparation

1. Warm SUV solution to room temperature on the bench top (at least 30 min).
2. Dilute SUVs in filtered PBS to a final concentration of 0.2 mg/mL.
3. For small wells (1-well Lab-Tek chamber), add 10 μL of the 0.2 mg/mL SUV solution to each well in the CultureWell gasket. For large wells (8-well Lab-Tek chamber), add 250 μL of the 0.2 mg/mL SUV solution to each well and mark the side of the 8-well chamber at the meniscus in each well using a permanent marker. This last step is to provide the user with a visual guide to ensure consistent volumes during subsequent incubation steps.
4. Incubate at room temperature in a humidity chamber for 45 min (*see Note 8*).
5. Wash the wells to remove unfused vesicles by continuous liquid exchange, using 4 mL PBS for large wells (Movie S2 in the online version at https://doi.org/10.1016/bs.mcb.2024.03.002) and 1 mL PBS for small wells (Movie S3 in the online version at https://doi.org/10.1016/bs.mcb.2024.03.002). Avoid the formation of bubbles, which can be detrimental to bilayer integrity.
6. Use a pipette to remove excess liquid, such that there is 250 μL remaining in the large wells (at the marked meniscus level) and 10 μL remaining in the small wells.
7. Block planar lipid bilayers with 100 μg/mL BSA. To do so, dilute BSA to 600 μg/mL in PBS. For large wells, add 50 μL of the 600 μg/mL BSA to each well. For small wells, add 2 μL of 600 μg/mL BSA to each well. Incubate for 1 h at room temperature or overnight at 4 °C to block exposed glass surfaces with BSA (*see Note 9*).
8. Use a pipette to remove 50 μL of liquid from each large well, or 2 μL of liquid from each small well.

9. Dilute streptavidin to $6\,\mu g/mL$ in PBS. Add $50\,\mu L$ to each large well, or $2\,\mu L$ to each small well, for a final concentration of $1\,\mu g/mL$. Incubate at room temperature for 20 min.

10. Wash the wells continuously with 4 mL PBS for large wells and 1 mL PBS for small wells to remove unbound streptavidin.

11. Add biotinylated DNA-antigen sensors at the desired concentration and incubate at room temperature for 10 min. We recommend a concentration range of 1 to 20 nM as a starting point, which can be tuned to achieve desired densities on the bilayer surface as described in Section 3.5.

12. Wash the wells with 4 mL PBS for large wells and 1 mL PBS for small wells to remove unbound sensors.

13. Check the quality of the bilayers on a fluorescence microscope, using 561 nm excitation (or similar) to visualize the Atto550 fluorophore. Good-quality bilayers are homogeneous and mobile. The mobility of the bilayers can be assessed by bleaching a small area of the bilayer and (i) monitoring fluorescence recovery by FRAP, or (ii) by tracking the lateral diffusion of individual sensors on the bilayer surface. If the surface is fluid, FRAP measurements will reveal that the fluorescence intensity recovers over time in the bleached region (Nair et al., 2011). Likewise, if the surface is fluid, single-particle tracking analysis of individual sensors (using the TrackMate plugin in Fiji (Tinevez et al., 2017)) will return a diffusion constant of $2-3\,\mu m^2/s$.

3.5 Lipid calibration standards

Antigen density impacts the B cell response and should be controlled and quantified for experiments. To do this, we prepare lipid bilayer standards that incorporate Liss Rhod PE, a fluorophore that is spectrally similar to the Atto550 on the DNA sensor. This method is adapted from Galush, Nye, and Groves (2008). We provide ImageJ macros to quantify antigen densities on bilayers, with a detailed description on our Github page: https://github.com/SpillaneLab/.

3.5.1 Generate a calibration curve

1. Prepare bulk lipid calibration standards made up of 4 mg/mL SUVs containing DOPC and Liss Rhodamine PE in varying proportions, yielding solutions with final amounts of Liss Rhodamine PE ranging from 0 to 0.06 mol% (Table 1). Follow the SUV preparation protocol described in Section 3.4.1.

Table 1 Volumes of different lipid stock solutions to be mixed together to create 2 mL SUV suspensions with the desired mol% of Liss Rhod PE and a final concentration of 4 mg/mL in PBS

Liss Rhod PE (mol%)	Liss Rhod PE (molec./μm^2)	μL of 25 mg/mL DOPC needed	μL of 1 mg/mL Liss Rhod PE needed
0	0	320	0
0.01	139	319.95	1.32
0.02	278	319.89	2.65
0.03	417	319.84	3.97
0.04	556	319.79	5.29
0.05	694	319.74	6.62
0.06	833	319.68	7.95

2. Prepare planar lipid bilayers in 8-well Lab-Tek chambers as described in Section 3.4.2.
3. Measure fluorescence intensities of the bulk lipid calibration standards using a fluorescence microscope, focusing on the bilayer plane. We recommend acquiring 36 images in different areas of each bilayer (Fig. 4A). Measure the power density at the sample using an optical power meter so that the imaging conditions can be replicated when determining antigen-DNA sensor densities (Section 3.5.3).
4. Generate a lipid calibration plot. To do this, take the mean fluorescence intensity value for each lipid calibration bilayer (averaged from 36 images) and plot these values against the surface density (molecules/μm^2) of Liss Rhodamine PE (estimate using the known mol% of Liss Rhodamine PE and a lipid footprint of 0.72 nm^2). The measured intensity should be linear with respect to Liss Rhodamine PE concentration (Fig. 4B).

3.5.2 Measure the scaling factor, F

Spectrally similar fluorophores, or the same fluorophore linked to different chemical moieties, have different absorption and emission characteristics (Galush et al., 2008). To take these differences into account, the intensity of the sample fluorophore (DNA sensor Atto550) must be calibrated to the bilayer standard fluorophore (Liss Rhod PE).

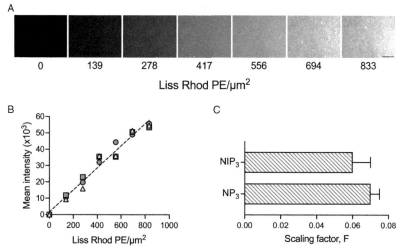

Fig. 4 Antigen-DNA sensor density calibration. (A) Representative images of planar lipid bilayer standards with increasing amounts of Liss Rhodamine PE. Scale bar, 10 μm. (B) Mean fluorescence intensity of bilayers from panel (A). Each data point represents the mean value from 36 different areas of the same bilayer. Different shapes/colors represent different independent experiments. The intensity increases linearly up to approximately 830 Liss Rhod PE molecules/μm². (C) Scaling factors for NIP₃- and NP₃-functionalized DNA sensors. Error bars represent the standard deviation of 64 (NIP₃-DNA) or 30 (NP₃-DNA) readings from one pair of DNA and vesicle samples. Imaging was performed with a 561 nm laser with power density of 7.4 W/cm² at the sample.

1. Prepare solutions of the sample fluorophore (Atto550) and lipid vesicle standard solution with the same molar concentration of dye. Add 250 μL of each solution to separate wells in an 8-well Lab-Tek chamber.
2. Measure the fluorescence intensity of the lipid vesicle solution ($I_{solu(lipid)}$) and the sample-fluorophore solution ($I_{solu(sample)}$) by defocusing the microscope into the solution, about 2 μm above the glass coverslip. The two samples should be measured with identical conditions (i.e., same laser power density, filter settings, exposure times, etc.).
3. Calculate the unitless value of F using Eq. (2):

$$F = \frac{I_{solu(sample)}}{I_{solu(lipid)}} \qquad (2)$$

4. Scaling factors can change depending on the chemical moiety to which the fluorophore is attached, and should be performed for each

sample–standard pair. Scaling factors for the DNA sensors described in Section 3.1 are given in Fig. 4C.

3.5.3 Determine the surface density of antigen-DNA sensors

1. After the antigen–DNA sensors have been loaded onto bilayers (Section 3.4.2), acquire 36 images of the Atto550 fluorophore in different areas with the same imaging settings used to generate the calibration curve in Section 3.5.1, maintaining focus on the bilayer surface using the microscope's autofocus unit.
2. In ImageJ, use the Get_mean_fluorescence_intensity macro to measure the mean fluorescence intensity of the 36 acquired images. The macro will output a .csv file containing the mean intensity value for each image. The mean of these values is the sample intensity, I_{sample}. Convert I_{sample} to the calibrated sample intensity, I_{cal}, using Eq. (3):

$$I_{cal} = \frac{I_{sample}}{F} \qquad (3)$$

3. Read the calibrated sample intensity (I_{cal}) from the calibration curve to determine the number of Atto550 molecules/μm^2 (Fig. 4B). This value is equal to the number of DNA sensors/μm^2.

3.6 Measuring antigen extraction by primary B cells at fixed timepoints

Primary B cells that express an NIP-specific BCR are isolated by negative selection from the spleens of B6.129P2(C)-Igh^{tm2Cgn}/J mice (B1-8i mice). Primary B cells should be prepared immediately before imaging, kept cold throughout (use ice-cold MACS buffer), and warmed to 37 °C before seeding onto bilayer substrates. Cells begin to internalize antigens from the sensors within approximately 5 min, so we recommend this as a minimum timepoint for cell fixation and analysis.

1. Filter 50 mL of MACS buffer through a 0.22 μm sterile syringe filter into a 50-mL Falcon tube, place the tube into a desiccator connected to a vacuum pump, and de-gas under vacuum until bubbling or "boiling" no longer occurs (at least 30 min). Store the buffer on ice until use. Degassed MACS buffer should be used for all subsequent cell isolation steps (including column equilibration, cell resuspension,

and cell elution) as the formation of air bubbles can clog the column, resulting in reduced yield and purity of B cells.

2. Mash the spleen through a 70 μm cell strainer using the plunger of a 2 mL syringe into a 50 mL falcon tube. Rinse the filter with 10 mL MACS buffer. Centrifuge for 5 min at $350 \times g$ and 4 °C.

3. Resuspend the cell pellet in 3 mL RBC lysis buffer and incubate for 4 min at room temperature to lyse red blood cells.

4. Top up the cell suspension to 10 mL with MACS buffer and centrifuge for 5 min at $350 \times g$ and 4 °C.

5. Resuspend cell pellet in 450 μL MACS buffer and add 50 μL anti-mouse CD43 (Ly-48) MicroBeads. Incubate on ice for 20 min (*see Note 10*).

6. Top up the cell suspension to 10 mL with MACS buffer and centrifuge for 5 min at $350 \times g$ and 4 °C.

7. During centrifugation, insert an LD column into a MidiMACS separator, apply 2 mL of fresh de-gassed MACS buffer, and let the buffer run through into a 15 mL Falcon collection tube. Discard effluent and change the collection tube.

8. Resuspend cell pellet in 500 μL ice-cold MACS buffer.

9. Load the cell suspension onto the LD column and collect the flow-through containing B cells into a 15 mL Falcon tube.

10. Wash the LD column with 2×1 mL MACS buffer. Collect the flow-through containing B cells into the same falcon tube (*see Note 11*).

11. Count the number of live B cells using a hemocytometer, using trypan blue staining to distinguish viable from nonviable cells.

12. Centrifuge cells for 5 min at $350 \times g$ and 4 °C.

13. Resuspend cell pellet to 5×10^6 cells/mL in warm full RPMI medium and culture in a 6-well multiwell plate at 37 °C with 5% CO_2. Allow the cells to recover in medium for 1 h before proceeding with experiments. We recommend using cells for experiments within 6 h of spleen isolation.

14. To prepare cells for imaging, remove 5×10^5 cells per large imaging well and centrifuge in a 1.5 mL microcentrifuge tube for 5 min at $350 \times g$ and 25 °C. If using small imaging wells, 5×10^5 cells is sufficient for 25 wells (20,000 cells per small well).

15. Resuspend cells in 50 μL per 1×10^6 cells in warm HBSS 0.1% BSA and incubate for 5 min at 37 °C by placing the microcentrifuge tube into an incubator or water bath.

16. Exchange imaging wells into warm HBSS 0.1% BSA, ensuring 250 μL buffer in each large well or 10 μL in each small well. Because the addition of BSA reduces the liquid surface tension, care must be taken to prevent overflow of the small wells.

17. Pipette cells into the wells containing antigen–DNA-loaded bilayers. Incubate at 37 °C for the desired timepoint to allow cells to extract and internalize antigens. We recommend timepoints within the range of 10–45 min.

18. Fix cells in 2.6% PFA by adding 50 μL 16% PFA to 250 μL (large well), or 2 μL 16% PFA to 10 μL (small well), and incubating at room temperature for 10 min. Gently wash fixed cells with HBSS 0.1% BSA by hand by filling wells and removing liquid carefully with a pipette, repeating several times to ensure at least 4× excess buffer exchange. After the last wash, restore the volumes to 250 μL (large wells) or 10 μL (small wells).

19. To block the binding of staining antibodies to B cell Fcγ receptors, add 12.5 μL normal mouse serum to each large well, or 0.5 μL to each small well, for a final concentration of 5% (*v/v*). Incubate for 30 min at room temperature. This is a crucial step to ensure specific labeling (*see Note 12*).

20. Label the cell membranes by adding 50 μL (to 250 μL large wells) or 2 μL (to 10 μL small wells) of 6 μg/mL anti-mouse B220 BV421 directly to the blocked wells, to achieve a final concentration of 1 μg/mL. Incubate at room temperature for 30 min.

21. Wash wells by hand with at least 4× excess buffer exchange and fix again as in step 18 to retain the labels over time.

22. Optional: Cells can be permeabilized and stained for intracellular markers. We recommend the FoxP3 Staining Kit from BD Biosciences.

23. Transfer the imaging chamber to the microscope stage for image acquisition. B cells attached to bilayers should be visible throughout the well. Acquire z-stacks covering a volume starting several planes below the bilayer and ending several planes above the B cells using a z-step size of 500 nm.

3.7 Quantitative image analysis

A common way to analyze the efficiency of B cell antigen internalization is to quantify the percentage of available antigen that B cells internalize from the immune synapse (Nowosad, Spillane, & Tolar, 2016; Spillane & Tolar, 2017). This can be determined by measuring the antigen fluorescence intensity in the cell (above the synapse plane), and dividing it by the total antigen fluorescence intensity in the cell and in the synapse. We provide a set of ImageJ macros and detailed instructions to measure the amount of antigen that cells extract and internalize from the bilayer substrates. These are available on our GitHub page: https://github.com/SpillaneLab/.

Briefly, the macros automatically correct images for background and inhomogeneous illumination, and then detect, segment, and crop individual cells from flatfielded z-stack images. The z-stack files are then re-sliced and summed to generate a side-view of each cell that shows the integrated antigen fluorescence both on the bilayer substrate and inside the cell. The user manually crops and saves two regions from each image: (i) cell + synapse (internalized + synaptic antigen) and (ii) cell (internalized antigen, starting typically $\sim 1.5\,\mu m$ above the synapse plane). In the specified regions, the macros identify the cell edge using the B220 membrane stain and quantify the integrated Atto647N intensity of antigen bound to the substrate and internalized by the cell. The integrated Atto647N intensities of total synaptic and internalized antigen for each cell (from the "cell + synapse" cropped image) and the total internalized antigen for each cell (from the "cell" cropped image) are saved in .csv files. The percent of antigen internalized from the synapse can then be calculated for each cell as

$$\%\text{Ag internalized} = \frac{\text{Integrated intensity}_{\text{Atto647N,cell}}}{\text{Integrated intensity}_{\text{Atto647N,cell+synapse}}} \times 100$$

This analysis approach offers a quick and simple method to assess the efficiency of antigen internalization and can be used to compare B cell responses to different stimuli (i.e., different antigen affinity or antigen density). If more detailed information is required (e.g., number or internalized clusters, or analysis of secondary stains), we direct readers to other image analysis tools for quantifying endocytosis (Morales-Navarrete et al., 2015; Nowosad and Tolar, 2017).

4. Notes

1. The reported rupture forces for short dsDNA molecules are based upon single-molecule force experiments that measured the level of constant force required to dissociate a DNA duplex within 2s

(Hatch, Danilowicz, Coljee, & Prentiss, 2008). However, dsDNA rupture forces have been shown to depend strongly on both loading rate (Cocco, Monasson, & Marko, 2001; Liu et al., 2023) and loading duration (Mosayebi, Louis, Doye, & Ouldridge, 2015), which both can change in the context of extracellular forces (Gohring et al., 2021). In our experiments, we assume that the loading rate and duration in the B cell synapse are the same for a given antigen affinity and density. This would mean that the "weak" dsDNA tethers will rupture at a lower force than the "strong" dsDNA tethers, leading to differences in the efficiency of antigen internalization.

2. The plastic Lab-tek chambers can be washed and re-used. Following an experiment, use tweezers to remove the coverslip and any silicone from the chamber, wash with copious amounts of ddH$_2$O, and sterilize with 70% ethanol. The chambers can then be re-used by attaching a freshly cleaned glass coverslip.

3. The silicone gaskets can be washed and re-used. Following an experiment, use tweezers to remove the strip of gaskets from the glass coverslip, wash with copious ddH$_2$O, rinse with absolute ethanol, and air dry. Store the cleaned gaskets in a sealed plastic bag. Before reapplying the silicone gasket strip to a fresh glass coverslip, remove any dust/debris from both sides of the silicone gasket using clear adhesive tape. The silicone gaskets can be re-used until they no longer form a good seal with the glass coverslip, typically three to four times.

4. The total reaction volume is 30.6 µL (4 oligos × 7.5 µL/oligo + 0.6 µL MgCl$_2$). If a larger reaction scale is required, perform multiple reactions of 30.6 µL in parallel rather than increase the reaction volume in a single PCR tube to ensure even temperature distribution. Perform desalting steps 2–6 separately for each 30.6 µL aliquot to ensure high sample purity.

5. The antigen-conjugated DNA probes should be thawed in single-use aliquots and protected from light to minimize photobleaching of the fluorophores. Once an aliquot is thawed for an experiment, any excess probe should be discarded. In our experience, the integrity of the annealed DNA probes is compromised with multiple freeze-thaw cycles.

6. The volumes of lipids used will change depending upon the concentration of the lipid stock solutions and the desired mol% of the lipid components. To calculate the required volumes, we use the supplementary spreadsheet provided by Nair et al. (2011).

7. If the lipid mixture does not become clear, or if the user does not have a bath sonicator, then SUVs can be prepared by mechanically extruding the lipid mixture through polycarbonate membranes with an 0.1 μm pore size using a mini-extruder. Mechanical extrusion gives better control than bath sonication over the vesicle size, which can lead to more reproducible bilayers. We use the mini-extruder set with holder/heating block from Avanti Polar Lipids. We recommend a total of 21 passes through the membrane to ensure a homogeneous lipid solution. The final extrusion should yield a clear solution.

8. To assemble a humidity chamber, take a clean, empty, lidded pipette tip box, remove the tip rack, fill the bottom compartment halfway with ddH$_2$O, and replace the tip rack. Place the chamber inside on a Kim-wipe and close the lid fully.

9. Planar lipid bilayers should be blocked with 100 μg/mL BSA prior to the addition of streptavidin. The blocking step prevents the formation of immobile streptavidin aggregates that can appear when streptavidin binds nonspecifically to glass that is exposed by bilayer defects, or when streptavidin forms 2D crystals at high concentrations (Reviakine & Brisson, 2001). Higher concentrations of BSA (e.g., 1 mg/mL) are not advised as it can lead to patches in the bilayer, as demonstrated by Nair et al. (2011).

10. The MicroBeads bind all CD43-expressing cells, which includes nearly all leukocytes except immature and mature resting B cells. The cells labeled with CD43 MicroBeads will be retained on the magnetic column, allowing untouched B cells to be collected in the flow-through fraction.

11. The LD columns can be washed and re-used. Following the isolation of unlabeled B cells, remove the column from the MidiMACS separator and flush out the magnetically labeled non-B cells by firmly pushing the plunger into the column. Wash immediately 3× with 3 mL of ddH$_2$O, followed 3× with 3 mL of absolute ethanol. Allow the column to air dry at room temperature by keeping the plunger loose at the top of the column. Cell numbers should always be counted after isolation, and if the cell numbers begin to drop then the LD column should be discarded. Eventually LD columns will rust, and these columns should be discarded.

12. B cells express high levels of Fcγ receptors, so the blocking step is crucial to minimize off-target labeling by the staining antibodies. Do not use normal mouse serum if the staining protocol requires an anti-mouse

secondary antibody. Instead, use normal serum from the species in which the secondary antibody was raised. As an alternative, purified anti-mouse CD16/32 mAb 2.4G2 (BD Biosciences) can be used at a concentration of 1 µg/mL to block Fcγ receptors.

5. Conclusions

The present protocol describes a method to assemble antigen-functionalized DNA tension sensors for the quantitative measurement of B cell antigen extraction in immune synapses formed with planar lipid bilayers. By changing antigen affinity and duplex stability, the relative strength of "tugging" and "tethering" bonds can be modified. This protocol can be adapted to investigate B cell antigen extraction from other substrates including planar lipid bilayers with different compositions and hydrogels (Spillane & Tolar, 2017). The sensors can also be used to measure mechanical forces in other cell types by attaching different ligands that engage the relevant cell surface receptors.

Acknowledgments

Figures were created in BioRender. This work was supported by a BBSRC research grant (BB/S007814/1), a BBSRC sLoLa (BB/V003518/1), and a Royal Society research grant (RGS\R2\180333) to K.M.S.

References

Albrecht, C., Blank, K., Lalic-Multhaler, M., Hirler, S., Mai, T., Gilbert, I., et al. (2003). DNA: A programmable force sensor. *Science, 301*(5631), 367–370.

Andrecka, J., Spillane, K. M., Ortega-Arroyo, J., & Kukura, P. (2013). Direct observation and control of supported lipid bilayer formation with interferometric scattering microscopy. *ACS Nano, 7*(12), 10662–10670.

Bajic, G., Yatime, L., Sim, R. B., Vorup-Jensen, T., & Andersen, G. R. (2013). Structural insight on the recognition of surface-bound opsonins by the integrin I domain of complement receptor 3. *Proceedings of the National Academy of Sciences of the United States of America, 110*(41), 16426–16431.

Batista, F. D., Iber, D., & Neuberger, M. S. (2001). B cells acquire antigen from target cells after synapse formation. *Nature, 411*(6836), 489–494.

Batista, F. D., & Neuberger, M. S. (1998). Affinity dependence of the B cell response to antigen: A threshold, a ceiling, and the importance of off-rate. *Immunity, 8*(6), 751–759.

Batista, F. D., & Neuberger, M. S. (2000). B cells extract and present immobilized antigen: Implications for affinity discrimination. *The EMBO Journal, 19*(4), 513–520.

Bruhns, P., Iannascoli, B., England, P., Mancardi, D. A., Fernandez, N., Jorieux, S., et al. (2009). Specificity and affinity of human Fcgamma receptors and their polymorphic variants for human IgG subclasses. *Blood, 113*(16), 3716–3725.

Cocco, S., Monasson, R., & Marko, J. F. (2001). Force and kinetic barriers to unzipping of the DNA double helix. *Proceedings of the National Academy of Sciences of the United States of America, 98*(15), 8608–8613.

Crotty, S. (2015). A brief history of T cell help to B cells. *Nature Reviews. Immunology, 15*(3), 185–189.

Dustin, M. L. (2016). Complement receptors in myeloid cell adhesion and phagocytosis. *Microbiology Spectrum, 4*(6).

Edelstein, A., Amodaj, N., Hoover, K., Vale, R., & Stuurman, N. (2010). Computer control of microscopes using μManager. *Current Protocols in Molecular Biology, 92*(1).

Galush, W. J., Nye, J. A., & Groves, J. T. (2008). Quantitative fluorescence microscopy using supported lipid bilayer standards. *Biophysical Journal, 95*(5), 2512–2519.

Gohring, J., Kellner, F., Schrangl, L., Platzer, R., Klotzsch, E., Stockinger, H., et al. (2021). Temporal analysis of T-cell receptor-imposed forces via quantitative single molecule FRET measurements. *Nature Communications, 12*(1), 2502.

Hatch, K., Danilowicz, C., Coljee, V., & Prentiss, M. (2008). Demonstration that the shear force required to separate short double-stranded DNA does not increase significantly with sequence length for sequences longer than 25 base pairs. *Physical Review. E, Statistical, Nonlinear, and Soft Matter Physics, 78*(1 Pt. 1), 011920.

Heesters, B. A., Chatterjee, P., Kim, Y. A., Gonzalez, S. F., Kuligowski, M. P., Kirchhausen, T., et al. (2013). Endocytosis and recycling of immune complexes by follicular dendritic cells enhances B cell antigen binding and activation. *Immunity, 38*(6), 1164–1175.

Iliopoulou, M., Bajur, A. T., McArthur, H. C. W., Gabai, M., Coyle, C., Ajao, F., et al. (2024). Extracellular matrix rigidity modulates physical properties of subcapsular sinus macrophage-B cell immune synapses. *Biophysical Journal.* https://doi.org/10.1016/j.bpj.2023.10.010.

Jiang, H., & Wang, S. (2023). Immune cells use active tugging forces to distinguish affinity and accelerate evolution. *Proceedings of the National Academy of Sciences of the United States of America, 120*(11), e2213067120.

Jung, Y., Wen, L., Altman, A., & Ley, K. (2021). CD45 pre-exclusion from the tips of T cell microvilli prior to antigen recognition. *Nature Communications, 12*(1), 3872.

Keppler, A., Gendreizig, S., Gronemeyer, T., Pick, H., Vogel, H., & Johnsson, K. (2003). A general method for the covalent labeling of fusion proteins with small molecules in vivo. *Nature Biotechnology, 21*(1), 86–89.

Liu, J., Le, S., Yao, M., Huang, W., Tio, Z., Zhou, Y., et al. (2023). Tension gauge tethers as tension threshold and duration sensors. *ACS Sensors, 8*(2), 704–711.

Los, G. V., Encell, L. P., McDougall, M. G., Hartzell, D. D., Karassina, N., Zimprich, C., et al. (2008). HaloTag: A novel protein labeling technology for cell imaging and protein analysis. *ACS Chemical Biology, 3*(6), 373–382.

Morales-Navarrete, H., Segovia-Miranda, F., Klukowski, P., Meyer, K., Nonaka, H., Marsico, G., et al. (2015). A versatile pipeline for the multi-scale digital reconstruction and quantitative analysis of 3D tissue architecture. *eLife, 4*, e11214.

Mosayebi, M., Louis, A. A., Doye, J. P., & Ouldridge, T. E. (2015). Force-induced rupture of a DNA duplex: From fundamentals to force sensors. *ACS Nano, 9*(12), 11993–12003.

Nair, P. M., Salaita, K., Petit, R. S., & Groves, J. T. (2011). Using patterned supported lipid membranes to investigate the role of receptor organization in intercellular signaling. *Nature Protocols, 6*(4), 523–539.

Natkanski, E., Lee, W. Y., Mistry, B., Casal, A., Molloy, J. E., & Tolar, P. (2013). B cells use mechanical energy to discriminate antigen affinities. *Science, 340*(6140), 1587–1590.

Nowosad, C. R., Spillane, K. M., & Tolar, P. (2016). Germinal center B cells recognize antigen through a specialized immune synapse architecture. *Nature Immunology, 17*(7), 870–877.

Nowosad, C. R., & Tolar, P. (2017). Plasma membrane sheets for studies of B cell antigen internalization from immune synapses. In C. T. Baldari, & M. L. Dustin (Eds.), *The immune synapse* (pp. 77–88). New York, NY: Humana Press.

Reviakine, I., & Brisson, A. (2001). Streptavidin 2D crystals on supported phospholipid bilayers: Toward constructing anchored phospholipid bilayers. *Langmuir, 17*, 8293–8299.

Saltukoglu, D., Ozdemir, B., Holtmannspotter, M., Reski, R., Piehler, J., Kurre, R., et al. (2023). Plasma membrane topography governs the 3D dynamic localization of IgM B cell antigen receptor clusters. *The EMBO Journal, 42*(4), e112030.

Sarrias, M. R., Franchini, S., Canziani, G., Argyropoulos, E., Moore, W. T., Sahu, A., et al. (2001). Kinetic analysis of the interactions of complement receptor 2 (CR2, CD21) with its ligands C3d, iC3b, and the EBV glycoprotein gp350/220. *The Journal of Immunology, 167*(3), 1490–1499.

Schindelin, J., Arganda-Carreras, I., Frise, E., Kaynig, V., Longair, M., Pietzsch, T., et al. (2012). Fiji: An open-source platform for biological-image analysis. *Nature Methods, 9*(7), 676–682.

Schmidt, A. G., Xu, H., Khan, A. R., O'Donnell, T., Khurana, S., King, L. R., et al. (2013). Preconfiguration of the antigen-binding site during affinity maturation of a broadly neutralizing influenza virus antibody. *Proceedings of the National Academy of Sciences of the United States of America, 110*(1), 264–269.

Schwickert, T. A., Victora, G. D., Fooksman, D. R., Kamphorst, A. O., Mugnier, M. R., Gitlin, A. D., et al. (2011). A dynamic T cell-limited checkpoint regulates affinity-dependent B cell entry into the germinal center. *The Journal of Experimental Medicine, 208*(6), 1243–1252.

Spillane, K. M., & Tolar, P. (2017). B cell antigen extraction is regulated by physical properties of antigen-presenting cells. *The Journal of Cell Biology, 216*(1), 217–230.

Spillane, K. M., & Tolar, P. (2018a). Mechanics of antigen extraction in the B cell synapse. *Molecular Immunology, 101*, 319–328.

Spillane, K. M., & Tolar, P. (2018b). DNA-based probes for measuring mechanical forces in cell-cell contacts: application to B cell antigen extraction from immune synapses. In C. Liu (Ed.), *B cell receptor signaling. 1707* (pp. 69–80). New York, NY: Springer New York.

Stayton, P. S., Freitag, S., Klumb, L. A., Chilkoti, A., Chu, V., Penzotti, J. E., et al. (1999). Streptavidin-biotin binding energetics. *Biomolecular Engineering, 16*(1–4), 39–44.

Suzuki, K., Grigorova, I., Phan, T. G., Kelly, L. M., & Cyster, J. G. (2009). Visualizing B cell capture of cognate antigen from follicular dendritic cells. *The Journal of Experimental Medicine, 206*(7), 1485–1493.

Tinevez, J. Y., Perry, N., Schindelin, J., Hoopes, G. M., Reynolds, G. D., Laplantine, E., et al. (2017). TrackMate: An open and extensible platform for single-particle tracking. *Methods, 115*, 80–90.

Victora, G. D., & Nussenzweig, M. C. (2022). Germinal centers. *Annual Review of Immunology, 40*, 413–442.

Wang, X., & Ha, T. (2013). Defining single molecular forces required to activate integrin and notch signaling. *Science, 340*(6135), 991–994.

Wedemayer, G. J., Patten, P. A., Wang, L. H., Schultz, P. G., & Stevens, R. C. (1997). Structural insights into the evolution of an antibody combining site. *Science, 276*(5319), 1665–1669.

Woodside, M. T., Behnke-Parks, W. M., Larizadeh, K., Travers, K., Herschlag, D., & Block, S. M. (2006). Nanomechanical measurements of the sequence-dependent folding landscapes of single nucleic acid hairpins. *Proceedings of the National Academy of Sciences of the United States of America, 103*(16), 6190–6195.

> CHAPTER SIX

Gauging antigen recognition by human primary T-cells featuring orthotopically exchanged TCRs of choice

Vanessa Mühlgrabner[a], Angelika Plach[a], Johannes Holler[a], Judith Leitner[b], Peter Steinberger[b], Loïc Dupré[c], Janett Göhring[a,*], and Johannes B. Huppa[a,*]

[a]Medical University of Vienna, Center for Pathophysiology, Infectiology and Immunology, Institute for Hygiene and Applied Immunology, Vienna, Austria
[b]Medical University of Vienna, Center for Pathophysiology, Infectiology and Immunology, Institute for Immunology, Vienna, Austria
[c]Medical University of Vienna, University Clinics for Dermatology, Vienna, Austria
*Corresponding authors: e-mail address: janett.goehring@meduniwien.ac.at; johannes.huppa@meduniwien.ac.at

Contents

1.	Introduction	128
2.	Material	131
3.	Methods	134
	3.1 Protein production	134
	3.2 UV-mediated peptide exchange	135
	3.3 Tetramer production	135
	3.4 Cell preparation and staining for single cell sorting	136
	3.5 TCR sequencing using Sanger sequencing	136
	3.6 CRISPR Cas9-mediated TCR exchange	140
	3.7 Preparation for ratiometric calcium imaging	142
	3.8 Determining antigen densities on SLBs	144
	3.9 Determining T-cell sensitivity via ratiometric calcium imaging	146
	3.10 TCR-specific ligand recruitment assay	147
4.	Notes	150
	4.1 Protein production	150
	4.2 T-cell stimulation	150
	4.3 Design of CRISPR guide RNAs	150
	4.4 TCR sequencing	150
	4.5 Imaging	151
5.	Conclusion	151
	Acknowledgments	152
	References	152
	Further reading	154

Methods in Cell Biology, Volume 193
ISSN 0091-679X
https://doi.org/10.1016/bs.mcb.2024.03.003

Copyright © 2025 Elsevier Inc.
All rights are reserved, including those
for text and data mining, AI training,
and similar technologies.

Abstract

Understanding human T-cell antigen recognition in health and disease is becoming increasingly instrumental for monitoring T-cell responses to pathogen challenge and for the rational design of T-cell-based therapies targeting cancer, autoimmunity and organ transplant rejection. Here we showcase a quantitative imaging platform which is based on the use of planar glass-supported lipid bilayers (SLBs). The latter are functionalized with antigen (peptide-loaded HLA) as adhesion and costimulatory molecules (ICAM-1, B7-1) to serve as surrogate antigen presenting cell for antigen recognition by T-cells, which are equipped with T-cell antigen receptors (TCRs) sequenced from antigen-specific patient T-cells. We outline in detail, how the experimental use of SLBs supports recoding and analysis of synaptic antigen engagement and calcium signaling at the single cell level in response to user-defined antigen densities for quantitative comparison.

1. Introduction

Single cell-TCR sequencing and CRISPR-Cas9-based TCR gene editing provide unique opportunities for deciphering the complexity of antigen-specific T-cell populations as they emerge in settings of infection, cancer, autoimmunity and transplant medicine (Han, Glanville, Hansmann, & Davis, 2014; Newell & Davis, 2014; Pauken et al., 2022). Once disease-relevant immunodominant epitopes have been identified, for example by combining the use of peptide/MHC (pMHC) multimers with flow cytometry, their cognate TCRs can be sequenced, reintroduced into primary human T-cells and tested for functionality. Linking clonotypic TCR genes to their corresponding epitopes in such fashion offers unprecedented insights into how T-cell immunity adapts in individual patients over time.

Of note, T-cell mediated detection of antigens underlies a high degree of plasticity as recognition outcomes largely depend on (i) the quality of a particular TCR-pMHC match as well as (ii) the differentiation and metabolic state of a given T-cell. T-cells are in fact capable of sensing the presence of even a single stimulatory pMHC on antigen presenting cells (APCs) or target cells (Huang et al., 2013; Irvine, Purbhoo, Krogsgaard, & Davis, 2002; Purbhoo, Irvine, Huppa, & Davis, 2004; Sykulev, Joo, Vturina, Tsomides, & Eisen, 1996), which renders them in principle a perfect therapeutic tool. However, in many physiological settings T-cell antigen sensitivities can be considerably reduced. This is in large part because the TCR-repertoire available for antigen recognition had been shaped by negative thymic selection to prevent autoimmunity. As a consequence, autoreactive T-cells typically require high densities of autoantigens for activation. Negative selection may furthermore limit the recognition of non-mutated tumor-associated antigens that are similar to self-antigens, let alone overexpressed self-antigens. De novo appearing neoantigens arising from tumor-specific mutations can induce potent antigen-specific T-cell responses as they are not limited by central tolerance (Schumacher & Schreiber, 2015). Although the number of theoretically predicted neoantigen epitopes is large, only a small fraction is recognized by T-cells (Strønen et al., 2016). The underlying mechanisms of this limitation are still under investigation but could be accounted to improper antigen processing and presentation, immune editing by tumor cells or tolerization of T-cells to neo-antigens (Gouttefangeas, Klein, & Maia, 2023; Schumacher & Schreiber, 2015; Strønen et al., 2016; Zamora, Crawford, & Thomas, 2018). While this notion constitutes at current a major limitation of T-cell-based cancer therapies, it also calls for patient-specific TCR-engineering, which may become, if productive biomedical and bioinformatic routines are streamlined, commonplace in the foreseeable future.

Critical for future success is the observation that differences as subtle as the absence or presence of a single methyl group within TCR-contacting peptide residues (e.g., threonine to serine mutation) can convert an agonist pMHC into a weak agonist pMHC with a more than 10-fold loss in affinity (Huppa et al., 2010). Vice versa, subtle changes within the peptide-contacting Complementarity Determining Regions (CDR) 3 largely affect T-cell antigen recognition. Interestingly, while CDR3s are on average 10 amino acids in length, only 3–4 residues within the CDR3s mediate peptide specificity (Glanville et al., 2017). This may in fact explain the high degree of cross-reactivity observed in TCRs, which likely expands on the one hand the universe of recognizable antigens, yet also increases risks for developing autoimmunity. Off-target toxicity due to cross-reactivity is a concern in the context of therapies that are based on the use of engineered TCRs with high affinity especially for self-antigens which would otherwise be excluded from the physiological TCR repertoire due to negative selection. (Cameron et al., 2013; Ishii et al., 2023). Hence, methodologies that are capable of precisely quantitating the capacity of any given TCR to convey T-cell recognition of antigens of interest, will be instrumental to the engineering and selection of high quality TCRs for therapeutic purposes in settings of cancer, infection/vaccination as well as autoimmunity and transplantation medicine through TCR-engineering of cytolytic T-cells (CTL), helper T-cells (T_H) regulatory T-cells (T_{reg}), respectively.

Here we showcase the identification of epitope-specific human SARS-CoV-2-specific T-cell clones from whole blood with the use of pMHC-multimers, pairwise sequencing of the cognate TCRα and TCRβ chains, their genetic reintroduction into primary human T-cells and (single molecule) imaging-based quantitation of their antigen detection capacity in the context of the immunological synapse (Fig. 1). We anticipate that the presented workflow will become instrumental for the identification and functional characterization of TCRs that turn out suitable for subsequent TCR-engineering and safety assessment in the context of T-cell-based therapeutic intervention in cancer, autoimmunity and transplant rejection.

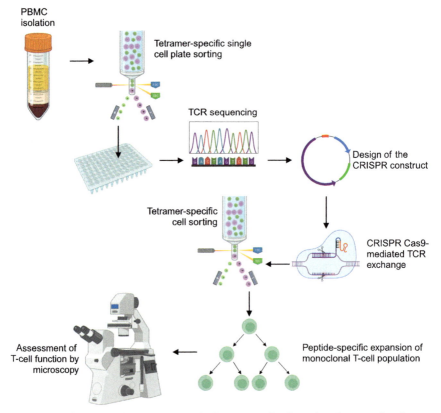

Fig. 1 Workflow describing the individual steps involved in identifying and isolating tetramer-specific TCRs from PBMCs. First, PBMCs are isolated from donor blood, single cell-sorted for T-cell markers and pMHC-tetramer specificity and subjected to TCR sequencing. Second, CRISPR-constructs are designed for replacing the endogenous TCRs with the TCRs identified via single cell sequencing. Following a second pMHC T-cell-based sorting step, cells are expanded and used for assessment of T-cell sensitivity and function via molecular imaging based on the use of protein-functionalized planar glass-supported lipid bilayers. *This figure was created with biorender.com.*

2. Material

Primary T-cells employed for the expression of individual TCRs of choice

1. Peripheral blood mononuclear cells (PBMCs): isolated from donor cartridges of the blood transfusion medicine department 4i of the general hospital of Vienna
2. K562 cell line expressing CD80 and HLA.A2
3. Lymphoprep (Stemcell, Cat no 07851)
4. Culture medium:
   a. For cell lines: RPMI 1640 medium containing L-glutamine (Invitrogen, Cat no 21875091) supplemented with 10% heat-inactivated fetal calf serum (FCS, Biowest), 100 U/mL penicillin/streptomycin (Gibco, Cat no 10378016), 2 mM L-Glutamine (Gibco, Cat no 25030024)
   b. For primary cells: RPMI 1640 medium containing L-glutamine supplemented with 10% human serum from male blood type AB negative donors (Pan Biotech, Cat no P30-2902), 2 mM Glutamax (Gibco, Cat no 35050061), 25 mM HEPES (Sigma, cat no H0887-100ML) and 50 μM β-mercaptoethanol (Gibco, Cat no 31350010).
5. Cytokines: Human IL-15 (Miltenyi, Cat no 130-095-762), Human IL-2 (Aldesleukin, Novartis).
6. Antibodies employed for T-cell activation: Ultra-LEAF purified anti-human CD3 (OKT3, Mouse IgG2a, kappa, Biolegend, Cat no B317326) and CD28 (T44, NA/LE, Clone: CD28.2, Isotype: Mouse C3H × BALB/c IgG1, κ, Cat no 555725)
7. Buffer used for T-cell sorting: 1×HBSS (Invitrogen, Cat no 14175129) supplemented with 1% FCS (Biowest) and 20 mM EDTA (Gibco, Cat no 15575020)
8. Antibodies employed for T-cell sorting: CD8 (OKT8, eFluor780, Invitrogen, Cat no 47-0086-42) and IP26 (TCRα/β AF488, Biolegend, Cat no B306712)

CRISPR-Cas9 mediated orthotopic TCR exchange

1. CRISPR guide RNAs, tracrRNAs, Alt-R S.p. Cas9 Nuclease V3 (Cat no 1081059) and Alt-R Cas9 Electroporation Enhancer (Cat no 1075916; all from Integrated DNA Technologies), Ribolock RNAse inhibitor (Thermo Scientific, Cat no EO0381).

Fig. 2 CRISPR construct comprised of left homology arm, P2A site, TCR β variable region (VDJ β), TCR β constant region (TRBC), T2A site, TCR α variable region (VDJ α), TCR α constant region (TRAC), Stop codon, bovine growth hormone polyadenylation signal (gGHpA), right homology arm.

2. SARS-CoV-2-specific TCR construct: This construct was designed based on obtained TCR sequences and ordered from Twist Biosciences as a CRISPR construct (including HDR arms) in a pet28b+ vector (see Fig. 2).
3. Human T-cell electroporation kit (Lonza, Cat no LONVPA-1002) and Amaxa Nucleofector II electroporator.

Proteins
1. pMHC molecules were expressed, refolded and purified in-house, as described in Section 3.
2. Regenerated cellulose Ultracel ultrafiltration disc 10 kDa cutoff (Merck, Cat no PLGC06210)
3. Amicon stirred cell (Merck, Cat no C3259)
4. Alexa Fluor 647 C2-maleimide (Invitrogen, Cat no A20347).
5. Tetramer formation was performed as described in the NIH T-cell facility protocol (https://tetramer.yerkes.emory.edu/support/protocols#10).
6. Biotin agarose (Sigma, Cat no B0519-5ML)
7. GST agarose (Thermo Scientific, Cat no 16100)
8. 365 nm UV lamp (Waveform lighting, 20W UV LED Food lamp, Cat no 77022.65)

TCR sequencing
1. Qiagen OneStep RT PCR kit (Qiagen, Cat no 210212)
2. Hot Start Taq polymerase and reaction buffer (Qiagen, Cat no 203203)
3. For phase 3 reactions (see protocol in Section 3.5 and Table 5) our in-house produced Taq polymerase was used.

Production of planar glass-supported lipid bilayers
1. Lipids
    1-palmitoyl-2-oleoyl-glycero-3-phosphocholine (POPC, Cat no 850457C)
    1,2-dioleoyl-*sn*-glycero-3-[(N-(5-amino-1-carboxypentyl) iminodiacetic acid)succinyl] (NiDGSNTA, Cat no 790404C)
    1,2-dioleoyl-*sn*-glycero-3-phosphoethanolamine-N-[methoxy (polyethylene glycol)-5000] (ammonium salt, PEG5000, Cat no 880230C)

Fura-2 AM ratiometric calcium imaging

1. 16-well Labtek chamber (Labtek, Cat no 178599PK)
2. Borosilicate glass slides ($22 \times 64\,mm^2$ no. 1.5 borosilicate; Menzel-Gläser)
3. Two component Picodent silicone (Picodent Twinduo, Cat no 13001001)
4. Fura-2 AM (Invitrogen, Cat no F1221)
5. The Inverted microscope (DMI4000; Leica Microsystems) is equipped with a $20\times$ objective (HC PL FLUOTAR $20\times/0.50$ PH2∞/0.17/D, Leica Microsystems) and a mercury lamp (EL6000; Leica Microsystems). A fast filter wheel containing 340/26 and 387/11 bandpass filters (Leica Microsystems) and a sCMOS Prime95b camera (Photometrix) are used for excitation. A Fura-2 emission filter (510/80) is used for the analysis of emission. Hardware components of the imaging system are operated by the open-source software Micromanager. All measurements were performed using a heating chamber (Leica Microsystems).
6. Fiji Image J
7. MATLAB (custom code by Janett Göhring, PhD based on Salles et al., 2013; Gao & Kilfoil, 2009, for data analysis details also refer to Hellmeier et al., 2021).

pMHC recruitment assay

1. The inverted microscope system (Eclipse Ti-E; Nikon) is equipped with a $100\times$ objective (Nikon SR APO TIRF $100\times$, NA = 1.49). An L6Cc laser combiner box (Oxxius) with diode lasers of the wavelengths 405 nm (LBX-405-180), 488 nm (LBX-488-200), 515 nm (LBX-515-150), 532 nm (LCX-532-500), 561 nm (LCX-561-250) and 640 nm (LCX-640-500) coupled into a polarization-maintaining fiber served as an excitation light source. To enable imaging in total internal reflection fluorescence (TIRF) mode, first a two-lens telescope focuses the laser beam onto the back focal plane of the objective. As a second step, the laser beam is moved into TIRF illumination position using a translatable stage (Axmann, Schütz, & Huppa, 2015). For IRM images, illumination with a xenon lamp (Lambda LS lamp, Sutter Instrument Company) is employed. For fluorescence detection, the EMCCD camera Andor iXon Ultra 897 is used. Hardware components are controlled by USB-3114 analog and digital output system (Measurement Computing) the Metamorph software (Molecular Devices). Measurements were performed using a Pecon heating chamber (Axmann et al., 2015).
2. Open-source software Fiji Image J and ThunderSTORM plugin (https://zitmen.github.io/thunderstorm/).

3. Methods
3.1 Protein production

For pMHC production, the light and heavy chain of the MHC-I molecule are expressed individually as inclusion bodies in BL21 *E. coli*. Human beta-2-microglobulin (β2m, UniProt: P61769) and the extracellular domains of HLA-A*02:01 (UniProt: P01892) which are C-terminally fused to an AviTag for biotinylation have been cloned into pHN1 and pET28b+ vectors, respectively (Gudipati et al., 2020).

a. Grow 200 mL o/n cultures of single BL21 colonies transformed with β2m or HLA-A*02:01 expression plasmids in LB medium supplemented with antibiotic (50 μg/mL Kanamycin or 100 μg/mL Ampicillin) shaking (180 rpm) at 37 °C.

b. The following day, dilute the o/n culture 1:1000 into 16 L LB expression media for HLA-A*02:01 heavy chain and 3 L LB expression media for β2m.

c. Monitor bacterial growth until OD600 reaches 0.5, then induce protein expression by addition of Isopropyl β-D-Thiogalactoside (IPTG) at a final concentration of 1 mM.

d. After 4 h of protein expression, harvest the cultures by centrifugation (8000g, 20 min, 4 °C).

e. Purify inclusion bodies by sonication of bacterial pellets and washing of insoluble inclusion bodies with detergent buffer (50 mM Tris (pH 8.0) and 200 mM NaCl buffer containing 2 mM EDTA, 1% deoxycholic acid, 1% Triton X-100 (all Merck)). Subsequent washing steps were performed in wash buffer I (50 mM Tris (pH 8.0) and 200 mM NaCl buffer containing 1% Triton X-100 (all Merck)) and wash buffer II (50 mM Tris (pH 8.0) and 200 mM NaCl buffer containing 1 mM EDTA (all Merck)). After washing, dissolve inclusion bodies in 6 M guanidine hydrochloride (Merck) and fully denature by addition of 5 mM DTT.

f. Refold MHC-I complexes in vitro in the presence of an ultraviolet light-cleavable peptide (KILGFVF[*ANP*]V, ANP = 3-amino-3-(2-nitro) phenyl-propionic acid) according to Garboczi, Hung, and Wiley (1992), Clements et al. (2002) and Toebes et al. (2006).

g. Dialyze refolding reaction 3 × 8 h against 10 L in 1× PBS in a Spectra por 12–14 kDa cut-off dialysis membrane (Spectrum, Cat no 132680). It is critical to remove Arginine by this dialysis step, as it can interfere with subsequent protein purification steps.

h. Concentrate refolding reaction (no higher than 2 mg/mL to avoid protein aggregation) by nitrogen pressure-based sample concentration using an Amicon stirred cell equipped with a 10 kDa regenerated cellulose ultrafiltration disc (both Merck) and purify by size exclusion chromatography (SEC, Superdex 200 Increase 10/300 GL, Cytiva).

i. Biotinylate protein monomers for 20 min at 30 °C and then o/n at 4 °C following instructions provided by Avidity.com (in particular, regarding reaction conditions for BirA biotin ligase). After biotinylation, the GST-tagged BirA enzyme can be removed by incubation with GST-agarose (Thermo Scientific).

j. Perform a further SEC run to separate protein aggregates (from the biotinylation process in low salt conditions) from residual biotin and protein monomers.

k. Confirm successful biotinylation by an electrophoretic mobility shift assay (EMSA):

    i. Mix tetravalent streptavidin with pMHC-I molecule at molar ratios of 4:1, 1:1 and 1:4 in 1× PBS.

    ii. Load reactions onto a 10% SDS gel together with pMHC-I and tetravalent streptavidin alone, as controls.

    iii. Stain the gel using Coomassie staining and confirm higher molecular weight shifts which represent pMHC-I-streptavidin complexes.

3.2 UV-mediated peptide exchange

a. Prepare a 96-well plate with low protein binding properties and add exchange peptides to pMHC-I complexes at a molar protein:peptide ratio of 1:20 (1.5 and 30 μM, respectively) in 1×PBS, here we use a SARS-CoV-2 spike protein derived peptide with the amino acid sequence YLQPRTFLL (S269–277).

b. Seal the 96 well plate with parafilm to reduce evaporation.

c. Place the 96 well plate underneath a 365 nm UV lamp (Waveform lighting) at a distance of ~10 cm.

d. Expose exchange reactions to 365 nm light for 2 h at 4 °C (cold room) and further incubation without UV-light for 16 h at 4 °C.

3.3 Tetramer production

a. Mix fluorophore-conjugated streptavidin with peptide exchanged HLA-A*02:01 proteins at a final molar ratio of 4:1 by adding 1/10 of the total volume of streptavidin to the pMHC-I complex every 10 min on ice.

b. After completion of the tetramerization process, add 3 μL of biotin-agarose (Sigma) per 100 μL tetramerization reaction and incubate for 20 min on ice.

c. Filter the reaction via 0.22 μm spin-filter tubes to remove biotin-agarose-bound streptavidin from the tetramer mixture.

3.4 Cell preparation and staining for single cell sorting

Peripheral blood was drawn from a vaccinated and convalescent SARS-CoV-2 seropositive and HLA-A*0201-positive donor.

PBMCs are then isolated by 30 min density gradient centrifugation (400g, with break off) at RT using Lymphoprep.

a. After isolation or thawing, rest PBMCs o/n in culture medium for cell lines (recipe see Section 2).

b. On the day of cell sorting, wash cells 1× in sorting buffer (recipe see Section 2).

c. Stain cells by adding 0.5 μg pMHC-tetramer per 1×10^6 cells in 100 μL sorting buffer for 25 min on ice. This concentration is adequate for staining T-cells via high quality TCR-pMHC pairs (e.g., such as SARS-CoV-2 specific T-cells) but needs to be verified for each pMHC-tetramer-based staining.

d. After tetramer incubation, directly add 1 μL CD8 (OKT8, eFluor780, Invitrogen) and 2 μL IP26 (TCRα/β AF488, Biolegend) per 1×10^6 cells for 30 min. *Note*: it is crucial to use the clone OKT8 when staining for CD8 expression. Clone SK7 inhibits tetramer binding.

e. Wash cells 3× with 3 mL sorting buffer.

f. Sort cells for expression of CD8 and αβTCR as well as a double-positive tetramer staining. Cells are either sorted in single-cell mode for Sanger sequencing methods or bulk sorted and further processed for 10xGenomics sequencing. In this chapter we cover Sanger sequencing.

3.5 TCR sequencing using Sanger sequencing

For detailed information refer to the published procedure (Han et al., 2014). Spray all reagents and materials (except for PCR primers & purchased RT/PCR kit components) with RNaseZap decontamination solution and UV-irradiate for 20 min prior to usage. Perform all pipetting and aliquoting steps in a laminar flow cabinet to avoid (RNase) contamination. All steps involving transportation of plates and buffers are done on dry ice or ice, respectively.

a. Prepare primer stocks for phase 1 and phase 2 PCRs:
   i. Dilute all primers to 100 μM, prepare aliquots and store at −20 °C.
   ii. The 74 phase 1 primers target different variable regions of the TCR alpha and beta locus and are all pooled for Phase 1 PCR (=P1V).
   iii. For phase 2 reactions all variable region primers are again pooled (=P2V). These primers contain a common constant sequence "CCAGGGTTTTCCCAGTCACGAC" 3′ of the variable TCR alpha/beta sequence. This sequence is the same for all primers, thus it is also used as the forward primer for phase 3 PCR.
   iv. In addition, for each variable domain primer mix (P1V and P2V), there are 2 corresponding constant domain TRAC/TRBC region primers Cα1 and Cβ1 for phase 1 and Cα2 and Cβ2 for phase 2 PCR, respectively. These constant domain primers are stored separately and only combined with the variable primer pool right before master mix preparations.
   v. Thus, for the final reactions, the variable primer pool and the constant alpha and constant beta primers are pooled at a ratio of 16:1:1 (e.g., 300 μL P1V + 18 μL Cα1 + 18 μL Cβ1 = Phase 1 primer mix)

b. Prepare 96-well semi-skirted PCR plates for sorting by adding 12 μL buffer (=2.4 μL 5× buffer diluted with 9.6 μL DNase/RNase free water) per well. Carefully seal plates with heat-stable PCR foil and store at −20 °C until sorting.

c. After sorting exactly one single antigen-specific T-cell per well, add 4 μL of PCR master mix for phase 1 amplification and properly seal the plate with a PCR foil. Perform phase 1 PCR either immediately or alternatively store plates at −80 °C. After completion of the PCR, the plate can be stored at −20 °C see Tables 1 and 2.

Table 1 Composition of phase 1 PCR/RT reactions.

Components	Amount per sample
Single cell sort reaction	12 μL
5× Buffer	0.8 μL
Phase 1 primer mix (100 μM)	0.8 μL
Enzyme mix	0.64 μL
dNTPs (10 mM)	0.64 μL
H_2O	1.36 μL

Table 2 Thermocycler program of phase 1 PCR/RT reactions.

Step	Temperature	Time
1. Reverse transcription	50 °C	36 min
2. Initial hot start Taq activation	95 °C	15 min
3. Denaturation	94 °C	30s
4. Annealing	62 °C	1 min
5. Extension	72°C	1 min
Repeat steps 3–5 for 25 cycles		
6. Final extension	72 °C	5 min
7. Hold	4 °C	∞

Table 3 Composition of phase 2 PCR reactions.

Components	Amount per sample
Product of phase 1 PCR reaction	1 µL
10× buffer	1.2 µL
dNTP (10 mM)	0.24 µL
HotStarTaq polymerase	0.1 µL
Phase 2 primer mix (100 µM)	0.6 µL
H_2O	8.89 µL

d. Thaw phase 1 PCR plate and use 1 µL of the PCR product as the template for the Phase 2 PCR reaction using nested primers. After completion of the second PCR the plate is stored at −20 °C (see Tables 3 and 4).

e. Confirm successful PCR amplifications by analyzing 1 µL of the PCR product on a 1% agarose gel electrophoresis. Expect a band at approximately 250 bp size.

f. As done previously, 1 µL product of the phase 2 PCR is employed as the template for phase 3 PCR (see Tables 5 and 6).

g. At this step it is possible to use specific barcoding primers or paired-end primers which are suitable for MISeq (Illumina) sequencing. We used conventional Sanger sequencing and thus primers which only amplify the PCR product from phase 2 PCR are sufficient.

h. After agarose gel evaluation, purify the final phase 3 PCR product using a PCR clean-up kit (Qiaquick PCR & gel clean-up kit, Qiagen) and send out for sequencing.

Table 4 Thermocycler program of phase 2 PCR reactions.

Step	Temperature	Time
1. Initial Hot Start Taq activation	95 °C	15 min
2. Denaturation	94 °C	30 s
3. Annealing	62 °C	1 min
4. Extension	72 °C	1 min
Repeat steps 2–4 for 28 cycles		
5. Final extension	72 °C	5 min
6. Hold	4 °C	∞

Table 5 Composition of phase 3 PCR reactions.
Phase 3 PCR reaction

Components	Amount per sample
Product of phase 2 PCR reaction	1 μL
10× buffer	5 μL
dNTP (10 mM)	1 μL
Taq polymerase	5 μL
Paired end or sequencing primer forward (100 μM)	1 μL
Paired end or sequencing primer reverse (100 μM)	1 μL
H_2O	41.5 μL

Table 6 Thermocycler program of phase 3 PCR reactions.

Step	Temperature	Time
1. Initial Taq activation	95 °C	5 min
2. Denaturation	95 °C	30 s
3. Annealing	65 °C	30 min
4. Extension	72 °C	1 min
Repeat steps 2–4 for 35 cycles		
5. Final extension	72 °C	5 min
6. Hold	4 °C	∞

i. Match TCR sequences with a database of human TCR variable domains using the IMGT V-Quest software tool (refer to https://www.imgt.org/IMGT_vquest/input).

j. In our case, the most abundant TCR sequence aligned best with the TCR α variable domain (TRAV) 12-1*01 (100.00%), TCR α variable joining domain (TRAJ) 27*01 (88.46%). We obtained the amino acid sequence "CVVNRGDKSTF" for the CDR3 region. For TCR β matches corresponded to TCR β variable (TRBV) domain 7-9*01 (100%), TCR β joining domain (TRBJ) 2-2*01 (100%) and resulted in the amino acid sequence "CALGDANTGELFF" for the CDR3 region. Since our donor is SARS-CoV-2 seropositive, vaccinated and convalescent we expect these T-cells to be derived from the CD8 memory pool. If this information is critical, we advise investigators to undertake a phenotypic analysis of the T-cell differentiation state via flow cytometry focusing on the expression levels of CD45RA and CCR7.

k. Use these sequences to design the complete construct for CRISPR Cas 9-mediated TCR exchange. DNA sequences of the SARS-CoV-2-specific TCR were cloned into a CRISPR cassette as described (Schober et al., 2019) and the construct ordered in a pet28b+ vector (Twist Biosciences). An example is provided in Fig. 2.

3.6 CRISPR Cas9-mediated TCR exchange

Background: The knockout of the endogenous TCR is mediated by TRAC and TRBC RNPs which create DNA double strand breaks within the constant regions of the TCR α and β chains. Upon transcription this results in a truncated version of TCR α and a frameshift and preliminary stop codon for TCR β. Thus, the endogenous TCR is no longer functional (for details see Schober et al., 2019).

TCR knock-in is achieved by providing a template for homology-directed repair (HDR) after the Cas9 double strand break which contains homology sequences 5′ and 3′ of the introduced TCR targeting the first exon of the endogenous TRAC locus. This leads to the sequence design depicted in Fig. 2:

a. Isolate/thaw PBMCs from HLA-A*02:01 negative donors and rest in RPMI with 10% human serum + 1% L-Glutamine + 1% Penicillin/Streptomycin and 50 U/mL human IL-2 overnight. Optionally, PBMCs can be enriched for CD8+ T-cells via the Magnisort human CD8+ T-cell enrichment kit (Thermo Scientific).

b. Stimulate PBMCs or T-cells with plate-bound OKT3 antibody (2.5 µg/mL) and soluble anti-CD28 antibody (1 µg/mL) in culture medium for primary cells (see Section 2) + 180 U/mL human IL-2 and 5 ng/mL IL-15 for 48 h prior to CRISPR Cas9-mediated TCR exchange.

c. Prepare RNP complexes for targeted TCR α- and β-chain double knock-out:

d. Anneal the trans-activating RNA (tracrRNA, 80 µM, Integrated DNA Technologies) with either CRISPR RNA (crRNA) for TRAC: 5'-AGAGTCTCTCAGCTGGTACA-3' (from Ren et al., 2017) or crRNA for TRBC1 and TRBC2: 5'-GGAGAATGACGAGTGG ACCC-3' for TRBC (both 80 µM, Integrated DNA Technologies) at 95 °C for 5 min.

e. Cool the annealed guide RNAs (gRNAs) to RT, then add 20 µM electroporation enhancer and 12 µM Cas9 (24 µM stock, both Integrated DNA Technologies) to form RNP complexes for 20 min at RT.

f. Mix 3 µL RNP complex for TRAC + 3 µL RNP complex for TRBC with 2 µg of template DNA per reaction in a 96 well U bottom plate.

g. Centrifuge up to 5×10^6 cells per CRISPR reaction at 350g 5 min for.

h. Resuspend the pellet in 100 µL electroporation buffer (82 µL buffer + 18 µL supplement 1, Human T-cell nucleofection kit, Lonza). *Note*: The cells should not remain in the buffer for more than 3 min as this reduces viability.

i. Mix cells briefly with the prepared RNP and template DNA mix in the 96-well plate and directly proceed with electroporation.

j. Transfer cells into an electroporation cuvette (Lonza).

k. Electroporate using the Amaxa Nucleofector II program T23.

l. Immediately after completion of electroporation add 1 mL of pre-warmed culture medium for primary cells + 180 U/mL human IL-2 to the cuvette and transfer cells into a 24-well plate.

m. After 10 days of further cultivation, sort cells for tetramer-specificity, CD8 and TCRα/β expression.

n. Tetramer-sorted T-cells are expanded by co-cultivation with peptide-pulsed K562 cells:
   a. Add 10 µM YLQ peptide (dissolved in DMSO) to K562s expressing CD80 and HLA.A2 and incubate for 2 h at 37 °C.
   b. Irradiate K562 cells with 80 Gray and co-culture with tetramer-sorted T-cells at an effector:target ratio of 1:3 in human T cell medium supplemented with 180 U IL-2/mL.
   c. Observe outgrowth and expansion of antigen-specific T-cells over 10–12 days.

3.7 Preparation for ratiometric calcium imaging

3.7.1 Vesicle preparation for the formation of planar glass-supported lipid bilayers (SLBs)

For imaging, glass-supported lipid bilayers serve as a surrogate antigen presenting cell.

All lipids are purchased from Avanti Polar Lipids (also distributed by Merck). Prepare lipid mixture for bilayer formation in chemical fume hood. You may follow hands-on instructions provided by Axmann et al. (2015).

a. In a 250 mL round bottom glass flask dissolve 1-palmitoyl-2-oleoyl-glycero-3-phosphocholine (POPC) with 1,2-dioleoyl-*sn*-glycero-3-[(N-(5-amino-1-carboxypentyl)iminodiacetic acid)succinyl](NiDGSNTA) and 1,2-dioleoyl-*sn*-glycero-3-phosphoethanolamine-N-[methoxy (polyethylene glycol)-5000] (ammonium salt, PEG5000) in chloroform and mix at a molar ratio 50:1:1.

b. Evaporate the chloroform either in a rotary evaporator or by constantly turning the open round bottom flask in your hand until most of the organic solvent is evaporated.

c. Attach the flask to vacuum overnight.

d. Degas PBS. Suspend the dried lipid mixture with the 250 mL round bottom flask in 10 mL degassed PBS.

e. Fill the flask with inert gas such as nitrogen or argon, close it with a stopper and seal the flask with autoclave tape.

f. Place the sealed flask into a bath sonicator and sonicate the lipid suspension at 120–170 W until it has turned clear. This takes between 30 and 60 min.

g. Monitor the progress of vesicle formation spectro-photometrically: Ensue that the absorption of undiluted emulsion in comparison to PBS remains constant at 234 nm (as an approximate indicator for the lipid concentration) yet should drop significantly at 550 nm (due to reduced light scattering via larger particles).

h. To pellet heavy non-unilamellar vesicles, which interfere with the formation of a contiguous SLB, pellet the vesicle suspension by centrifugation, at 37,000g for 1 h at 25 °C and then for 8 h at 43,000g, 4 °C.

i. Filter the supernatant through a syringe filter with a pore diameter of 0.2 μm.

j. Measure the OD at 234 and 550 nm to monitor the clarity of the vesicle preparation and the amount of lipid left.

k. Store the vesicles at 4 °C under argon or nitrogen. Optionally, use vesicle suspension for bilayer preparation for several months.

3.7.2 Preparation of glass slides for spotting SLBs

a. Use 24×50 mm #1.5 glass slides (e.g., from VWR).
b. Prepare glass slides by cleaning them with plasma for 15 min (Zepto, Diener Electronic).
c. Attach a 16-well Labtek chamber (VWR) onto the glass slide using 2 component Picodent silicone (Picodent Twinduo) glue and allow to harden for 15 min at 37 °C.
d. Form a lipid bilayer on the cleaned glass surface by adding 100 µL lipid solution per well.
e. After a 15 min incubation step at RT rinse each well with 13 mL of $1 \times$ PBS (RT).

3.7.3 Preparation of proteins used for SLB-functionalization

a. For SLB-based experiments HLA.A2/peptide complexes are C-terminally fused to a poly-HisTag (12× histidine) to enable binding of the protein to NiDGS NTA head groups within the lipid bilayer. All expression and refolding steps are performed as described for Avi-tagged proteins (as described above).
b. The dialyzed refolding reaction is directly loaded onto a HisTrap FF column (1 mL, Cytiva) in running buffer ($1 \times$ PBS + 20 mM imidazole). *Note*: Thorough dialysis is essential to remove Arginine, which interferes with Nickel binding.
c. Elute bound protein from the NiNTA column by applying elution buffer (PBS + 300 mM imidazole) and collect in 0.5 mL fractions.
d. Measure protein concentration in elutions by A280 absorbance measurements (Nanodrop).
e. Pool fractions containing highest protein yields and run via SEC.
f. Protein monomers are site-specifically conjugated by a free cysteine on the N-terminus of β2m with Alexa Fluor 647 (Invitrogen) using maleimide chemistry according to manufacturer's instructions (Manual Thiol-Reactive Probes, Invitrogen).
g. Purify labeled proteins by an additional SEC run (here: S75) to remove unbound dye.
h. Human ICAM-1–12×His and B7.1–12×His are produced in Hi5 insect cells (BTI-Tn-5B1-4, Thermo Fisher Scientific) infected with viral particles of the Baculovirus Expression system according to Gudipati et al. (2020). Alternatively, they can be purchased from Sino Biological, Inc. or other vendors.

i. Functionalize SLBs by 1 h incubation with fluorescence-labeled HLA. A2/peptide, unlabeled ICAM-1 and unlabeled B7.1. Titrate antigen at densities ranging from 0.001 to 1000 molecules μm^{-2}. Verify antigen densities via TIRF microscopy:

3.8 Determining antigen densities on SLBs (details see Axmann et al., 2015)

a. *Note*: Single molecule intensities are approximated by measuring single fluorescence images in TIRF-mode and correlating it to bulk fluorescence measurements rather than employing per se single molecule resolution imaging techniques such as STORM or STED.

b. First, acquire 10–30 individual images of all well with varying densities.

c. For determination of single molecule intensities, acquire at least 50 individual images, each at a different location within the well. For this, choose an antigen density on the SLB where single-molecules can be easily distinguished from one another, and point-spread functions of events do not overlap.

d. Click Plugins > ThunderSTORM > Run analysis

e. Type in values for camera setup: Pixel size = 160 nm, Photoelectrics per A/D count = 15.3, Quantum efficiency = 0.1, EM gain = 300.

f. Use settings as summarized in Table 7.

Table 7 Software settings for detection of single molecules using ThunderSTORM.

Image filtering	
Filter	Wavelet filter (B–Spline)
B–Spline order	3
B–Spline scale	2.0
Approximate localization of molecules	
Method	Local maximum
Peak intensity threshold	2*std. (Wave.F1)
Connectivity	8-neighborhood
Sub-pixel localization of molecules	
Method	PSF: Integrated Gaussian
Fitting radius (px)	3
Fitting method	Maximum likelihood
Initial sigma (px)	1.6
Multi-emitter fitting analysis	Disabled

g. Plot obtained results as histogram: Transform units of intensity and offset values from photons to ADU by right-clicking and selecting "ADU". Then click "Plot histogram": Parameter: intensity, Bins: 100, Range: 0–10,000. Note single-molecule intensity and offset values (see Fig. 3).
h. Next, measure intensity values of all other densities by drawing a ROI around the laser profile and clicking "m" (see Fig. 3).
i. Average the 10–50 individual positions on the bilayer.
j. Subtract camera background values (see Fig. 3).

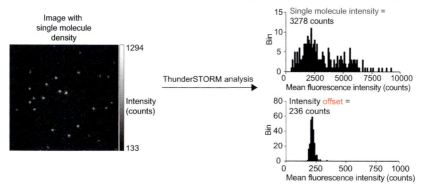

Fig. 3 Determination of antigen density on SLBs. (1) Offset and single molecule intensities are determined by ThunderSTORM analysis on images of SLB samples with single molecule densities. (2) Intensity measurements of antigen titrations are corrected for camera background (3), correlated with single-molecule intensities (4) and finally converted into molecules μm^{-2} by employing microscopy-setup specific conversion factors (5).

k. The number of molecules μm^{-2} can be derived by camera background corrected MFI values by single molecule intensity values (obtained by ThunderSTORM plugin measurements in Fiji Image J, see Fig. 3).

l. Apply a camera-specific correction factor for pixel size to μm^2 conversion and divide obtained value by the area of measurement (see Fig. 3).

3.9 Determining T-cell sensitivity via ratiometric calcium imaging

a. Add $4\,\mu L$ of Fura-2 AM (Invitrogen, dissolved in $50\,\mu L$ DMSO) per 1×10^6 T-cells and incubate for 15 min at 37 °C in 1 mL culture medium.

b. Wash cells in 10 mL imaging buffer ($1 \times$ HBSS + 5% ovalbumin + 2 mM calcium chloride + 2 mM magnesium chloride, $0.22\,\mu m$ filtered).

c. Pellet cells by centrifugation at $350g$ and resuspended at a density of 8000 cells/μL.

d. Spot 80,000 cells per imaging well.

e. Cells are imaged by alternating 340 and 380 nm lamp-based excitation every 15 s for 20 min at 37 °C. Upon T-cell activation, calcium is initially released from ER stores to result in capacitive entry of extracellular calcium. An increase in intracellular calcium levels mediates in an increase in the ratio of Fura-2 emissions following 340 and 380 nm excitation.

f. Antigen sensitivities are determined via the T-cell calcium response toward SLB-titrated antigen. The percentages of activating cells for each antigen density are determined using an in-house calcium analysis tool (devised by Janett Göhring, based on Salles et al., 2013). In brief, cells that are in contact with SLBs for at least 10 min are tracked using a particle tracking algorithm (as described in Gao & Kilfoil, 2009) and their mean Fura-2 ratio is derived from 340/380 ratio images (generated with Fiji Image J). For normalization of the Fura-2 ratio, T-cells confronted with SLBs featuring only ICAM-1 and B7.1 are used as a negative control, and high densities of activating ligand are used as the positive control.

g. Cells were classified as "activating" if their Fura-2 intensity value was higher than the threshold for 80% of the entire trace. Accordingly, "non-activating" cells showed Fura-2 intensity values below the threshold for 80% of the trace. If none of the two definitions apply, the cell is considered as "oscillatory." The percentages of activating T cells for each condition are fitted by a 3-parameter dose-response curve using the following equation: $Y = Bottom + X * \frac{Top-Bottom}{EC50 + X}$ and displayed as the fraction of activated T-cells as a function of antigen density (also see Hellmeier et al., 2021).

3.10 TCR-specific ligand recruitment assay

a. Functionalize SLBs by 1 h incubation with unlabeled ICAM and B7.1 and varying densities of HLA.A2/peptide-AF647 ranging from 0.001 to 1000 molecules μm^{-2}.

b. Measure initial density on the SLBs without cells (see Fig. 3) as described for density measurements above and according to Axmann et al. (2015).

c. Wash antigen-specific T-cells in 5 mL imaging buffer. Resuspend T-cells in imaging buffer at a density of 2000 cells/μL.

d. Spot 5 μL of the cell suspension per imaging well.

e. Allow cells to adhere to the SLBs for 5 min prior to imaging.

f. Confirm proper adhesion and spreading of the T-cells to the SLBs by Interference Reflection Microscopy (IRM).

g. To assess pMHC recruitment, acquire an AF647 fluorescence image using 640 nm TIRF illumination (20 ms illumination, 0.1 kW cm^{-2}, ET700/75 bandpass filter).

h. Cells are imaged for 5–10 min and pMHC recruitment is analyzed by intensity measurements in the open-source software Fiji Image J.

i. Fig. 4 provides a detailed description on the analysis of pMHC recruitment by T-cells:

j. Open Image J (Fiji) and determine mean intensity values (MFI, Fig. 4).

k. Determine initial ligand densities on SLBs without cells present on the bilayer as described in Section 3.8 and Axmann et al. (2015).

l. Measure fluorescence intensity under the synapse (Fig. 4).

m. Subtract camera background values (Fig. 4).

n. The number of molecules μm^{-2} can be derived by camera background corrected MFI values by single molecule intensity values (obtained by ThunderSTORM plugin measurements in Fiji Image J, Fig. 4).

o. Apply a camera-specific correction factor for pixel size to μm^2 conversion and divide obtained value by the area of measurement (Fig. 4).

p. pMHC recruitment can be calculated by subtracting initial ligand densities on SLBs from ligand densities upon T-cell encounter (to obtain the total number of recruited molecules). Alternatively, ligand densities upon T-cell engagement can be divided by initial ligand densities on SLBs (to obtain fold increase of initial density values).

q. Plot pMHC recruitment as fold increase of initial density vs initial number of ligands on SLBs (see Fig. 5).

1. Determine area of synapse:
Draw ROI around the cell and click "m" (=3060)

2. The same way, measure pMHC fluorescence intensity (Int Dens) within ROI (=7248966) and initial pMHC fluorescence intensity (=1187108, mean of 10-30 individual positions on bilayer)

3. Substract camera background (= offset):
Cell: 7248966 - (3060* 236) = 6526806
Initial density: 1187108 - (3060* 236) = 464948

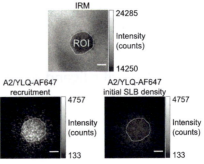

4. Divide background-corrected fluorescence intensity by single molecule intensity = total number of molecules within ROI.
Cell: 6526806 / 3278 = 1991
Initial density: 464948 / 3278 = 142

5. Convert fluorescence intensity to number of molecules per µm² by multiplication with a camera-specific conversion factor: 1 µm² = 39 pixel
Cell: 1991 / 3060 * 39 = 25 molecules µm^{-2}
Initial density: 142 / 3060 * 39 = 1.8 molecules µm^{-2}

6. Substract inital pMHC density from pMHC fluorescence intensity within cell synapse to determine number of recruited ligands within ROI.
Recruited ligands: 1991 - 142 = 1849 molecules within the synapse; 23.2 molecules µm^{-2};
14- fold increase versus initial density

7. Plot number of recruited ligands against initial density.

8. For low densities single molecules need to be tracked by Thunderstorm:
Cell: 10 molecules within ROI; area = 5345. This equals 0.073 molecules µm^{-2}
Initial density: 429 molecules detected in 75 frames = 5.72 molecules / frame; area = 40000;
This equals 0.75 molecules per ROI or 0.006 mol µm^{-2}
Recruited ligands: 10 - 0.75 = 9.25 molecules within the synapse; 0.067 molecules µm^{-2}
12- fold increase versus initial density

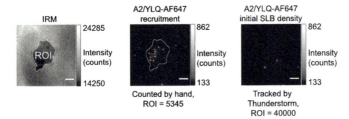

Fig. 4 Determination of antigen-specific pMHC recruitment under the synapse of cells on SLBs with varying densities of ligand. Initial ligand densities on the SLBs are calculated and used for correcting cellular recruitment intensity values. Note that for single-molecule intensities pMHC molecules need to be determined via ThunderSTORM or by manual counting. Scale bar = 5 µm.

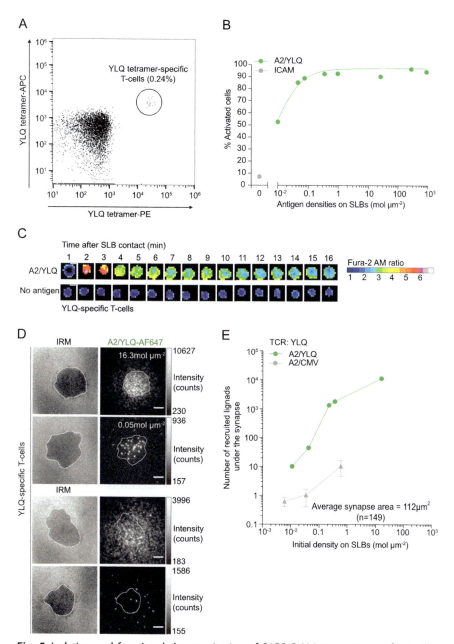

Fig. 5 Isolation and functional characterization of SARS-CoV-2-tetramer-specific T-cells. (A) Sorting strategy to arrive at tetramer-specific T-cells. Tetramer double-positive populations (black circle) were sorted. Pre-gated on CD8 positive, αβTCR positive T-cells, overlaid with the unstained control (gray). (B) T-cell antigen sensitivity measured by Fura-2 AM ratiometric calcium imaging. T-cells respond to antigen in a dose-dependent manner. No antigen (ICAM-1) control is represented by the gray datapoint. (C) Tracking the Fura-2 AM response of an individual T-cell over time with and without presence of antigen on SLBs. Images were acquired every 15 s. Scale bar = 3 μm. (D) Antigen recruitment underneath the T-cell synapse with (left panel) and without (right panel) specificity for the provided antigen. Scale bar = 5 μm. (E) Antigen recruitment plotted as fold increase of initial antigen density on SLBs.

4. Notes

4.1 Protein production

Efficiencies of pMHC refolding and biotinylation vary depending on the peptide provided for refolding reactions as well as on the HLA allele employed and tags C-terminally linked to the HLA heavy chain. In general, we expect higher protein production yields for Avi-tagged proteins than for His-tagged proteins. The protocols described in this chapter are optimized for HLA.A2 and need at times to be optimized for other the HLA alleles and types.

4.2 T-cell stimulation

CRISPR Cas9-mediated TCR exchange success rates become substantially reduced if T-cells are not properly activated (i.e., if T-cells fail to form visibly larger cell clusters and to consume medium). We successfully evaluated plate-bound OKT3/CD28 stimulation as well as commercial alternatives (Immunocult, Stemcell). Other activation procedures will need to be verified.

4.3 Design of CRISPR guide RNAs

When using online tools to design guide RNAs, always keep on-target and off-target scores as well as the position within the gene that is targeted for the double-strand break in mind. We recommend testing two to three guide RNAs followed by evaluation of knock-out efficiencies.

Also, not all TCRs inserted via the described CRISPR Cas9 approach are functionally expressed on the T-cell surface and enable T-cell effector functions. Again, designing and ordering two to three TCR constructs may turn out advantageous.

4.4 TCR sequencing

Approximately 10% of $\alpha\beta$ T-cells express two TCR α chains (Heath et al., 1995) and 1% expresses two TCR β chains (Balomenos et al., 1995) due to TCR gene rearrangement. The presence of such dual TCR-expressing T-cells can challenge TCR sequence analysis when employing conventional Sanger sequencing. One drawback of this method is its bias in primers that are chosen for PCR amplifications. Not all variants of the TCR variable regions which naturally exist are amplified equally well (due to sequence alterations) by the described primer pool. This bears the possibility of losing weakly amplified variable regions. In this regard, alternative sequencing methods such as 10× Genomics sequencing may prove superior.

4.5 Imaging

It is essential to spot T-cells as a single-cell suspension and at densities of 2000 cells/μL for pMHC recruitment or 8000 cells/μL for calcium imaging, respectively. These densities ensure that antigen recruitment is not hindered by spatial constrictions and that T-cells can be tracked as individual entities by the calcium analysis software. Also, the quality of the tool-aided calcium analysis relies on the quality of the negative control (ICAM-1 only on SLBs) which serves to define the threshold considered to be crossed for T-cell activation. We therefore recommend recording one backup "ICAM-1 only" sample per antigen titration series.

5. Conclusion

Here, we describe the isolation of SARS-CoV-2 specific, pMHC tetramer-sorted T-cells from peripheral blood of a seropositive donor. We determine the sequences of their cognate TCRα and β chains, design plasmids for CRISPR-Cas9-mediated TCR exchange and introduce the sequenced TCR into primary human T-cells. Subsequent imaging-based methods characterize the functionality of T-cells equipped with the discovered TCR with respect to antigen sensitivity and recruitment modalities. Of note, the employed microscopy methods are not generating images with single molecule resolution per se but acquire TIRF-images of single fluorescence intensities. However, super-resolution imaging such as stochastic optical reconstruction microscopy (STORM) or stimulated emission depletion (STED) microscopy could also be applied to further describe the molecular organization of immune synapses formed by these engineered T-cells (Sajman & Sherman, 2023). Apart from antigen sensitivity and recognition—as addressed in this chapter—the binding kinetics of TCR-pMHC interaction itself are indicative for the quality of the ensuing immune response and can be measured as previously described using a single molecule fluorescence resonance energy transfer (FRET)-based imaging approach (Brameshuber et al., 2018; Huppa et al., 2010). Furthermore, the importance of mechanical forces for the discrimination of agonistic from non-agonistic pMHC molecules has been proposed and could be instrumental for further characterizing these isolated TCR-pMHC interactions by DNA- or FRET-based force sensors or molecular traction force microscopy (Göhring et al., 2021; Issler, Colin-York, & Fritzsche, 2023; Zhang, Ge, Zhu, et al., 2014). This comprehensive characterization of sequenced TCRs can aid the selection of suitable candidates for TCR-based immunotherapies. Importantly, future studies employing engineered T-cells should also address safety aspects such as TCR cross-reactivity studies using peptide libraries, in silico cross-reactivity prediction tools or 2D and 3D human cell cultures (Gouttefangeas et al., 2023; Ishii et al., 2023; Sanderson et al., 2019).

Acknowledgments

This study was supported by funds of the Austrian Science Fund (FWF P 34118-B to V.M., L.D. and J.B.H., FWF P-32307-B to J.G. and FWF-P32411 to J.L. and P.S.).

References

Axmann, M., Schütz, G. J., & Huppa, J. B. (2015). Single molecule fluorescence microscopy on planar supported bilayers. *Journal of Visualized Experiments, 105*, e53158.

Balomenos, D., Balderas, R. S., Mulvany, K. P., Kaye, J., Kono, D. H., & Theofilopoulos, A. N. (1995). Incomplete T cell receptor V beta allelic exclusion and dual V beta-expressing cells. *Journal of Immunology, 155*(7), 3308–3312.

Brameshuber, M., Kellner, F., Rossboth, B. K., Ta, H., Alge, K., Sevcsik, E., et al. (2018). Monomeric TCRs drive T cell antigen recognition. *Nature Immunology, 19*(5), 487–496.

Cameron, B. J., Gerry, A. B., Dukes, J., Harper, J. V., Kannan, V., Bianchi, F. C., et al. (2013). Identification of a Titin-derived HLA-A1-presented peptide as a cross-reactive target for engineered MAGE A3-directed T cells. *Science Translational Medicine, 5*(197), 197ra103.

Clements, C. S., Kjer-Nielsen, L., MacDonald, W. A., Brooks, A. G., Purcell, A. W., McCluskey, J., et al. (2002). The production, purification and crystallization of a soluble heterodimeric form of a highly selected T-cell receptor in its unliganded and liganded state. *Acta Crystallographica Section D, 58*(12), 2131–2134.

Gao, Y., & Kilfoil, M. L. (2009). Accurate detection and complete tracking of large populations of features in three dimensions. *Optics Express, 17*, 4685–4704.

Garboczi, D. N., Hung, D. T., & Wiley, D. C. (1992). HLA-A2-peptide complexes: Refolding and crystallization of molecules expressed in *Escherichia coli* and complexed with single antigenic peptides. *Proceedings of the National Academy of Sciences of the United States of America, 89*, 3429–3433.

Glanville, J., Huang, H., Nau, A., Hatton, O., Wagar, L. E., Rubelt, F., et al. (2017). Identifying specificity groups in the T cell receptor repertoire. *Nature, 547*, 94–98.

Göhring, J., Kellner, F., Schrangl, L., Platzer, R., Klotzsch, E., Stockinger, H., et al. (2021). Temporal analysis of T-cell receptor-imposed forces via quantitative single molecule FRET measurements. *Nature Communications, 12*(1), 2502.

Gouttefangeas, C., Klein, R., & Maia, A. (2023). The good and the bad of T cell cross-reactivity: Challenges and opportunities for novel therapeutics in autoimmunity and cancer. *Frontiers in Immunology, 14*, 1212546.

Gudipati, V., Rydzek, J., Doel-Perez, I., Dos Reis Gonçalves, V., Scharf, L., Königsberger, S., et al. (2020). Inefficient CAR-proximal signaling blunts antigen sensitivity. *Nature Immunology, 21*, 848–856.

Han, A., Glanville, J., Hansmann, L., & Davis, M. M. (2014). Linking T-cell receptor sequence to functional phenotype at the single-cell level. *Nature Biotechnology, 32*, 684–692.

Heath, W. R., Carbone, F. R., Bertolino, P., Kelly, J., Cose, S., & Miller, J. F. (1995). Expression of two T cell receptor alpha chains on the surface of normal murine T cells. *European Journal of Immunology, 25*(6), 1617–1623.

Hellmeier, J., Platzer, R., Eklund, A. S., Schlichthaerle, T., Karner, A., Motsch, V., et al. (2021). DNA origami demonstrate the unique stimulatory power of single pMHCs as T cell antigens. *Proceedings of the National Academy of Sciences of the United States of America, 118*(4), e201687118.

Huang, J., Brameshuber, M., Zeng, X., Xie, J., Li, Q. J., Chien, Y. H., et al. (2013). A single peptide-major histocompatibility complex ligand triggers digital cytokine secretion in CD4(+) T cells. *Immunity, 39*(5), 846–857.

Huppa, J. B., Axmann, M., Mörtelmaier, M. A., Lillemeier, B. F., Newell, E. W., Brameshuber, M., et al. (2010). TCR-peptide-MHC interactions in situ show accelerated kinetics and increased affinity. *Nature, 463*(7283), 963–967.

Irvine, D. J., Purbhoo, M. A., Krogsgaard, M., & Davis, M. M. (2002). Direct observation of ligand recognition by T cells. *Nature, 419*(6909), 845–849.

Ishii, K., Davies, J. S., Sinkoe, A. L., Nguyen, K. A., Norberg, S. M., McIntosh, C. P., et al. (2023). Multi-tiered approach to detect autoimmune cross-reactivity of therapeutic T cell receptors. *Science Advances, 9*(30), eadg9845.

Issler, M., Colin-York, H., & Fritzsche, M. (2023). Quantifying immune cell force generation using traction force microscopy. *Methods in Molecular Biology, 2654*, 363–373.

Newell, E. W., & Davis, M. M. (2014). Beyond model antigens: high-dimensional methods for the analysis of antigen-specific T cells. *Nature Biotechnology, 32*, 648–692.

Pauken, K. E., Lagattuta, K. A., Lu, B. Y., Lucca, L. E., Daud, A. I., Hafler, D. A., et al. (2022). TCR-sequencing in cancer and autoimmunity: Barcodes and beyond. *Trends in Immunology, 43*(3), 180–194.

Purbhoo, M. A., Irvine, D. J., Huppa, J. B., & Davis, M. M. (2004). T cell killing does not require the formation of a stable mature immunological synapse. *Nature Immunology, 5*(5), 524–530.

Ren, J., Liu, X., Fang, C., Jiang, S., June, C. H., & Zhao, Y. (2017). Multiplexed genome editing to generate universal CAR T cells resistant to PD1 inhibition. *Clinical Cancer Research, 23*(9), 2255–2266.

Sajman, J., & Sherman, E. (2023). High- and super-resolution imaging of cell-cell interfaces. *Methods in Molecular Biology, 2654*, 149–158.

Salles, A., Billaudeau, C., Sergé, A., Bernard, A.-M., Phéliopt, M.-C., Bertaux, N., et al. (2013). Barcoding T cell calcium response diversity with methods for automated and accurate analysis of cell signals (MAAACS). *PLoS Computational Biology, 9*, e1003245.

Sanderson, J. P., Crowley, D. J., Wiedermann, G. E., Quinn, L. L., Crossland, K. L., Tunbridge, H. M., et al. (2019). Preclinical evaluation of an affinity-enhanced MAGE-A4-specific T-cell receptor for adoptive T-cell therapy. *Oncoimmunology, 9*(1), 1682381.

Schober, K., Müller, T. R., Gökmen, F., Grassmann, S., Effenberger, M., Poltorak, M., et al. (2019). Orthotopic replacement of T-cell receptor α- and β-chains with preservation of near-physiological T-cell function. *Nature Biomedical Engineering, 3*, 974–984.

Schumacher, T. N., & Schreiber, R. D. (2015). Neoantigens in cancer immunotherapy. *Science, 348*, 69–74.

Strønen, E., Toebes, M., Kelderman, S., van Buuren, M. M., Yang, W., van Rooij, N., et al. (2016). Targeting of cancer neoantigens with donor-derived T cell receptor repertoires. *Science, 352*(6291), 1337–1341.

Sykulev, Y., Joo, M., Vturina, I., Tsomides, T. J., & Eisen, H. N. (1996). Evidence that a single peptide-MHC complex on a target cell can elicit a cytolytic T cell response. *Immunity, 4*(6), 565–571.

Toebes, M., Coccoris, M., Bins, A., Rodenko, B., Gomez, R., Nieuwkoop, N. J., et al. (2006). Design and use of conditional MHC class I ligands. *Nature Medicine, 12*(2), 246–251.

Zamora, A. E., Crawford, J. C., & Thomas, P. G. (2018). Hitting the target: How T cells detect and eliminate tumors. *Journal of Immunology*, *200*(2), 392–399.

Zhang, Y., Ge, C., Zhu, C., et al. (2014). DNA-based digital tension probes reveal integrin forces during early cell adhesion. *Nature Communications*, *5*, 5167.

Further reading

Platzer, R., Hellmeier, J., Göhring, J., Perez, I. D., Schatzlmaier, P., Bodner, C., et al. (2023). Monomeric agonist peptide/MHCII complexes activate T-cells in an autonomous fashion. *EMBO Reports*, *28*, e57842.

CHAPTER SEVEN

Measuring interaction kinetics between T cells and their target tumor cells with optical tweezers

Edison Gerena[a,†], Sophie Goyard[b,†], Nicolas Inacio[a,b], Jerko Ljubetic[c], Amandine Schneider[c], Sinan Haliyo[a], and Thierry Rose[b,c,*]

[a]Institut des Systèmes Intelligents et de Robotique (ISIR), Sorbonne Université, CNRS, Paris, France
[b]Institut Pasteur, Université Paris Cité, Diagnostic Test Innovation & Development Core Facility, Paris, France
[c]Institut Pasteur, Université Paris Cité, INSERM-U1224, Unité Biologie Cellulaire des Lymphocytes, Ligue Nationale Contre le Cancer, Équipe Labellisée Ligue-2018, Paris, France
[*]Corresponding author: e-mail address: thierry.rose@pasteur.fr

Contents

1. Introduction	156
1.1 Synapse formation ignited by the CD3 engagement in the T cell contact to target cells	156
1.2 Mechanobiology toolbox for measuring cell pair interaction kinetics	157
1.3 Measuring interaction kinetics between T cell and P815 using optical tweezers	159
2. Material	161
2.1 Disposables	161
2.2 Reagents	161
2.3 Cells	162
2.4 Equipment	162
2.5 Software	162
3. Methods	163
3.1 Cell culture (8 days)	163
3.2 Glass-bottom Petri dish preparation (1 day)	163
3.3 Addition of the T cells and setting their motion control with the optical tweezer	164
3.4 Repetitive contacts of the same area of the two cells	165
3.5 Repetitive contacts of the same area of the T cell and different spots around P815	165
3.6 Rolling T cell all around P815 until immobilization	166
3.7 Force data analysis	167
4. Results	168
5. Concluding remarks	169

[†] These authors contributed equally to this work.

Methods in Cell Biology, Volume 193
ISSN 0091-679X
https://doi.org/10.1016/bs.mcb.2024.09.002

Copyright © 2025 Elsevier Inc.
All rights are reserved, including those
for text and data mining, AI training,
and similar technologies.

6. Perspectives	171
7. Notes	171
Acknowledgments	172
Declaration of interest	172
References	172

Abstract

T cell adhesion kinetics are a powerful indicator of target cell recognition during the cell-cell exploration process and formation of the immunological synapse facilitating cell communication and activation through specific intercellular molecular interactions. Various techniques have been used to document these binding kinetics, which foreshadow the dynamics of immunological synapse formation. Here, optical tweezers were used for studying at the level of single cells, the adhesion kinetics of human leukemia T lymphocyte cell line (CEM) to mouse mast cell line (P815) used as a tumor cell model. The P815 FcγRII receptors were saturated with the mouse anti-human CD3ε immunoglobulin G (OKT3) for initiating the T cell-P815 interaction through the engagement of the T cell CD3 nucleating the TCR complex formation structuring the synapse. Methods were developed to assess the time required to turn a contact between a T cell and a tumor cell into a stable interaction, and thus initiate the synapse formation. Single T cells were manipulated with the optical tweezers while the tumor cells were adhered to the glass surface under culture conditions. Three adhesions scenario were investigated by exerting either repetitive contacts engaging the same area of the two cells, repetitive contacts engaging the same area of the T cell but different areas on the tumor cell surface, or rolling the T cell over the tumor cell surface. With these methods, we observed that the median time of contact of CEM on P815 decreased in the presence of anti-CD3 OKT3 from 46 s to 1.3 s and the median rolling distance decreased from 50 μm to 1.8 μm prior the T cell immobilization. T cell adhesion speed assays can be used for measuring their lack of early response, identifying molecules involved in cell adhesion, or screening potential modulators.

The techniques and quantitative methods, described here for studying T cell/target cell interaction based on manipulations using optical tweezers, can be generalized to all types of immunological or virological synapses as between T cell/dendritic cell, cytotoxic T cell/target, T cell/macrophage, T cell/B cell, NK cell/target, immune cell/infected cell and others.

1. Introduction

1.1 Synapse formation ignited by the CD3 engagement in the T cell contact to target cells

The occurrence of immunological synapse between T lymphocytes and target cells is initiated by the T cell receptor (TCR) engagement with a presenting antigen at the target cell surface along random cell–cell contacts. This engagement results in the reorganization of both the actin and microtubule cytoskeleton driving T cell integrins reinforcing the adhesion force at the cell–cell contact site and triggers an array of signaling contributing to the organization and development of the synapse structure (Mastrogiovanni, Juzans, Alcover, & Di Bartolo, 2020). Polarity regulators have been described for controlling immunological synapse formation dynamics and organizing cell polarity such as Dlg1 and Apc (Adenomatous polyposis coli) (Aguera-Gonzalez et al., 2017; Juzans et al., 2020; Lasserre et al., 2010).

The early stages following a contact event and leading to an adhesion or an escape are critical in the immune response. The adhesion/escape decision time should be shorter than the contact event but tightly time-controlled to avoid blind adhesion.

To illustrate a method to investigate the early steps between contact and synapse formation, we have chosen the cell pairing model between a T lymphoblast cell line (CEM) isolated from the peripheral blood of a child suffering of acute lymphoblastoid leukemia (Foley et al., 1965) and the mast cell line (P815) isolated from a mouse with mastocyma (Gajewski, Markiewicz, & Uyttenhove, 2001). The P815 cells express the receptor FcγRII of immunoglobulin (Ig) G constant domain (Fc). A murine monoclonal anti-human CD3 antibody (IgG2a isotype) from mouse (OKT3) has been added to P815 and binds its cognate receptor FcγRII at the surface of cells, presenting anti-CD3 domain to cell contact (Landegren, Andersson, & Wigzell, 1984). CD3 is a complex of four transmembrane proteins (γ, δ and 2 ε) associated to the two TCR sub-units (α and β) at the surface of mature T cells and constitutive at the surface of CEM cells. Binding of OKT3 to CD3ε triggers the assembly of the TCR/CD3 complex and initiates an intracellular signaling pathway resulting in cellular activation and proliferation (Norman, 1995).

1.2 Mechanobiology toolbox for measuring cell pair interaction kinetics

The physical manipulation of single immune cells has become an approach of interest in immunology to document and understand the mechanical process involved in immune cell binding from its first contact to a target, its surface exploration to its tight adhesion and the formation of the synapse. Different techniques have been explored for working at the scale of a single cell, using traditional contact approaches as micropipettes (Hogan, Babataheri, Hwang, Barakat, & Husson, 2015; Shao & Hochmuth, 1996), atomic force spectroscopy (Benoit, Gabriel, Gerisch, & Gaub, 2000; Lehenkari & Horton, 1999; Li, Redick, Erickson, & Moy, 2003) as well as "non-contact" approaches involving optical (Ashkin, Dziedzic, & Yamane, 1987; Favre-Bulle & Scott,

2022; Schaffer, Norrelykke, & Howard, 2007; Wang et al., 2011) and magnetic (Gosse & Croquette, 2002) forces used as tweezers. Others non-contact techniques are allowing the analysis at the single cell level using electric force fields, acoustic waves (acoustic force spectroscopy; Kamsma et al., 2018; Sitters et al., 2015), dragging flow (laminar shear flow; Benoliel, Capo, Mege, & Bongrand, 1994; Juzans et al., 2020; Mastrogiovanni, Di Bartolo, & Alcover, 2022) but forces are exerted on all objects present in the force field. If magnetic tweezers require the cell labeling with paramagnetic particles or beads at their surface (Gosse & Croquette, 2002), optical tweezers do not and are one of the most widespread techniques for single cell manipulation (Favre-Bulle & Scott, 2022). An optical tweezer is produced by a near infra-red laser beam (1070 nm) whose convergence is controlled to create, if not a focal point, at least a constriction of the order of a micrometer (Fig. 1A) (Ashkin et al., 1987; Nieminen, Knöner, Heckenberg, & Rubinsztein-Dunlop, 2007; van Mameren, Wuite, & Heller, 2011). This constriction traps any object of higher density than its direct surroundings, such as a bead, a vesicle or a small cell in an aqueous solution, or a membrane, nucleus or organelle in the cytoplasm of a cell. The higher the density contrast, the stronger the force exerted and

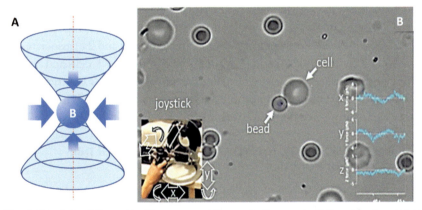

Fig. 1 Principle of the optical tweezer and its use in the manipulation of micrometric objects. (A) The bead B is trapped by the focused laser beam constriction, the optical trap, with the axial and lateral gradient forces figured by blue arrows. (B) The joystick is controlling x, y and z translational and rotational movements of the optical trap and the thumb and index are controlling a trigger driving the distance between two optical tweezers that can be used to clamp objects. The deviation between the center of the tweezer application and the real-time position of the manipulated object is converted to force along the 3 axes vs time as shown on the graph on the right. These forces are fed back to the joystick providing a real-time sensing of any motion resistance.

therefore the trap (Shaffer et al., 2007). The deviation between the center of the trap and the object is converted to force (Jahnel, Behrndt, Jannasch, Schäffer, & Grill, 2011). The position of the laser beam is fixed, but a three-axis joystick coordinates the movement of the microscope stage in x, y and z, controlled by piezoelectric motors with nanometric precision (Fig. 1B). The beam can be split into several beams using a split mirror. In this way, several objects can be manipulated simultaneously, or several grippers can be applied to different points on the same object to control its x, y and z translational and rotational movements along these three axes with the appropriate joystick.

Optical tweezer systems are marketed by several companies as add-ons or integrated into inverted microscopes. The lab-built setting described here uses high-speed image tracking in an event-based camera with nanometric resolution of movements in the three directions (Gerena, Legendre, Vitry, Regnier, & Haliyo, 2020; Yin, Gerena, Pacoret, Haliyo, & Régnier, 2017). This camera enables real-time measurement of the distance and orientation of the deviation of the trapped object's position from the center of the constriction. The greater the force applied to the object, the greater the deviation with an elastic spring law. As shown in Fig. 1B, the deviation is translated into force on the three axes x, y and z. This real-time force sensing allows the implementation of a haptic device providing transparent force feedback. The joystick used in this setting provides a force sensing of optical tweezer feedback force along the three axes with high bandwidth (up to 10 kHz) and low latency (Gerena et al., 2020; Gerena & Haliyo, 2023). The apparatus can sense forces in the piconewton range, and upscale these to the users' hands to let them operate with dexterity, and mechanically probe bio-samples' mechanical characteristics. This force feedback enables the user to feel the roughness of a surface or the movements of a living cell membrane.

Optical tweezers offer a significant advancement in studying interaction of immune cells with molecular or cell targets (Chen et al., 2007; Simpson, Bowden, Höök, & Anvari, 2002) for documenting and understanding cellular adhesion and synapse formation and dynamics along the immune response.

1.3 Measuring interaction kinetics between T cell and P815 using optical tweezers

In this study, cells are observed using a custom-made inverted transmission microscope with high-speed camera. For trapping micrometric objects, a near-infrared laser minimizing phototoxicity is piloted using a seven-degree of freedom joystick actuating high-speed laser-deflection generated by a galvanometer and a deformable mirror (Gerena et al., 2019). The 3D motion of the focal spot is obtained by the synchronization of the orientation of the galvanometer mirror and focusing or defocusing the deformable mirror. A 3D-nano stage is controlled by a piezo motor. The nano-stage moves the Petri dish while trapped T cells remain fixed. Multiple traps are created by sequentially moving the focal spot between different positions. This time-sharing of the laser beam on several spots is made possible by the short response time of the galvanometer and the deformable mirror in a real-time control framework. This design allows the creation of numerous independent optical traps within a volume of approximately $70 \times 50 \times 9\,\mu m^3$ with a bandwidth of up to 200 Hz. Tweezer force actuations are limited to 500 pN on T cells because of their low difference of cell density with the media. The optical tweezers were therefore not used to assess the force required to break the interaction between T cells and P815. We are taking advantage of the ease of manipulation offered by this technique to measure the time required to build an interaction that will require more than 100 pN to be broken. This corresponds to the force received by a surface-bound T lymphocyte in a local flow speed of 3 mm/s. This allows to reduce the laser power to the minimum for limiting localized photoactivation and formation of toxic free radicals.

Three approaches, illustrated in the Fig. 2, have been developed for assaying the time necessary for turning a contact of a T cell with a tumor cell into a stable interaction, initiating the synapse formation. Single T cells were manipulated with the tweezers while the tumor cells were adhered to the glass surface under culture conditions. First the number of repetitive contacts (0.2 s every second) were counted prior immobilization by moving

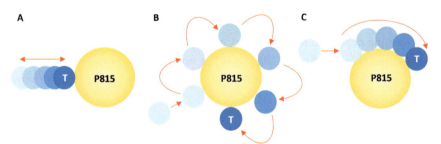

Fig. 2 Adhesion kinetic assays by repetitive contacts at the same spot (A), different spots (B) or by rolling the T cell (C) over the P815 cell. The P815 cells are adhered and immobilized on the glass surface of the Petri dish. The T cell is moved by a single optical tweezer applied to the center of the cell.

back and forth the T cell, engaging the same area of the two cells. Second the repetitive contacts (0.2 s every second) were counted engaging the same area of the T cell but different areas on the tumor cell surface. Third, the T cells were rolled over the tumor cell surface until immobilization, the time and the rolling distance from contact to immobilization were reported.

2. Material
2.1 Disposables
1. 10 μL, 200 μL, 1000 μL tips (Sorenson)
2. Glass bottom Petri dishes (MatTek, 10/35 mm, Ref# P35G-1.5-10-C)
3. Sterile microcentrifuge tubes (1.5 mL, Eppendorf Ref# 3810).
4. Petri dishes for cell culture (TPP, Sigma-Aldrich, Ref# Z707678-840EA)

2.2 Reagents
1. PBS, phosphate buffered saline (Gibco, Thermo Fisher Scientific, Ref# 14190).
2. Anti-CD3 antibody (Invitrogen, clone: OKT3, 16-0037-85)
3. Isotype control (IgG Armenian Hamster, eBioscience 17151252)
4. 0.05% Trypsin-EDTA 1× (Gibco, Thermo Fisher Scientific, Ref# 25300-054)
5. EDTA, Ethylenediaminetetraacetic acid (Sigma-Aldrich, Ref# E9884)
6. Ethanol (Sigma-Aldrich, Ref# 100983)
7. High-glucose (4.5 g/L) Dulbecco's Modified Essential Medium (DMEM).
8. Roswell-Park Memorial Institute medium (RPMI) 1640 medium with GlutaMAX-I and Phenol Red (Gibco, Thermo Fisher Scientific, Ref# 61870).
9. RPMI 1640 medium, without phenol red (Gibco, Thermo Fisher Scientific, Ref# 11835).
10. HEPES, N-(2-Hydroxyethyl)piperazine-N0-(2-ethanesulfonic acid) (Life Technologies, Ref# 15630-056).
11. Penicillin-streptomycin (Gibco, Thermo Fisher Scientific, Ref# 15140-122).
12. Fetal bovine serum (FBS, Dominique Dutscher, Ref# SV30160.03).
13. Poly-L-lysine 0.1% (Sigma, Ref# P8920)
14. CFSE CellTrace (CarboxyFluorescein diacetate Succinimidyl Ester; Thermo Fisher Scientific, Ref# C34554)

2.3 Cells

1. CEM T lymphoblast cell line isolated from a peripheral blood of a child with acute lymphoblastoid leukemia (Foley et al., 1965) distributed by ATCC (#CRL-2265)
2. P815 mast cell line isolated from a mouse with mastocyma (Gajewski et al., 2001) distributed by ATCC (#TIB-64).

2.4 Equipment

1. Cell culture incubator allowing standard cell culture conditions in a humidified atmosphere (37 °C, 5% CO_2)
2. Standard bench top centrifuge (Eppendorf, Ref# 5810R)
3. Sterile cell culture laminar flow hood safety level II
4. Custom-made inverted transmission microscope
   a. High-speed CMOS camera (Basler, Ahrensburg, Germany, 659 × 494 pixels)
   b. Oil immersion objective 40×, NA 1.3 (Olympus UPlanFLN 3)
   c. Nanometric controlled piezo-motorized stage
   d. Temperature and CO_2-controlled chamber (Stage-top incubator system T, LCI)
5. Optical tweezer
   a. Near-infrared laser (1070 nm, YLR-10-LP-Y12, IPG Photonics)
   b. High-speed laser-deflection generated by a galvanometer (GVS002, Thorlabs, Newton, NJ, USA)
   c. Deformable mirror (PTT111 DM, Iris AO, Berkeley, CA, USA)
   d. Event-based camera (Gen3S VGA-CD, Prophesee)
   e. Seven-degree of freedom (DoF) joystick (Omega.7, ForceDimension) three-DoF are for translations, three-DoF are for rotations and a last DoF is given by a gripper under the index and thumb fingers of the user for clamping single objects or to drive two objects closer or in contact.
6. Monocanal pipettes 0.1–10, 10–200, 100–1000 µL (Thermo Fisher, Finnpipette F1)
7. Dry-bath with aluminum block to thermostate cells in tubes or Petri dishes outside the incubator (Bioblock, Thermolyne, Ref# 92607)

2.5 Software

1. ImageJ (Schneider, Rasband, & Eliceiri, 2012) or Fiji (Schindelin et al., 2012) for image analysis.

2. Excel (Microsoft) or Prism (GraphPad) for data analysis.
3. An inlab-coded program for real-time control framework of the laser beam implemented on a Real-Time kernel (Xenomai).

3. Methods
3.1 Cell culture (8 days)

1. Non adherent P815 cells were thawed at room temperature from frozen vials (10% DMEM, HEPES 10 mM, 80% FBS, 10% DMSO, stored in nitrogen tank) then were grown in DMEM, HEPES 10 mM, 10% FBS, 1% (v/v) penicillin-streptomycin at 37 °C, CO_2 5% for a week in 75 cm^2 flask. Dead cells were removed in the flask supernatant. Seeding density of approx. 1×10^5 cells/mL was used for new flask of P815 cells. To ensure continued exponential growth, the cell density should be maintained between 1×10^5 and 1×10^6 cells/mL. After a week of cell grown, cells were ready to use.
2. Leukemia CEM T cell line was also thawed at room temperature from frozen vials (10% RPMI, HEPES 10 mM, 80% FBS, 10% DMSO, stored in nitrogen tank) and then cultured in 25 cm^2 flask of RPMI 1640 medium supplemented with 10% fetal bovine serum (FBS) and 1% (v/v) penicillin-streptomycin (Foley et al., 1965). After a week of cell grown, cells were ready to use.

3.2 Glass-bottom Petri dish preparation (1 day)

1. Load 200 μL of poly-lysine 1% on the glass bottom Petri dishes (MatTek, 10/35 mm). Incubate for 10 min then discard and wash twice with 200 μL of PBS. Prewarm the Petri dishes for 15 min at 37 °C in the incubator.
2. After a week of cell grown, centrifuge 1 mL of 10^5 P815 cells. Discard the supernatant. Resuspend cells in 10 mL of DMEM without FBS. Centrifuge for 5 min at 1000g. Resuspend P815 cells in 1 mL of DMEM, 10 mM HEPES with 1 μL of CFSE then gently mix by aspiration and expulsion. After 20 min of incubation at room temperature cells, centrifuge 5 min at 1000g. Resuspend cells in 10 mL of DMEM, 10 mM HEPES. Centrifuge cells 5 min at 1000g. Discard the supernatant. Resuspend P815 cells in 1 mL of DMEM, 10 mM HEPES with 10% FBS. Incubate cells at 37 °C for 1 h before use.
3. Load 10^4 P815 cells on the prewarmed poly-lysine-treated glass bottom Petri dishes in 100 μL of DMEM HEPES 10 mM. Incubated cells in dish

for 15 min at 37 °C, CO_2, 5% in humidified atmosphere. Ideal density is 100 cells per mm^2.
4. Check adhesion of cells using transmission light microscopy. Discard unbound cells with uncolored RPMI, HEPES 10 mM, FBS 10% in the absence or presence of antibody anti-OKT3 (0.1 μg/mL final concentration) incubated for 15 min at 37 °C.
5. See Notes 1 and 2 for cell adhesion and 3 and 4 for cell density.

3.3 Addition of the T cells and setting their motion control with the optical tweezer

1. Dilute 50 μL of the T cell suspension in 1 mL of uncolored RPMI HEPES 10 mM in a 1.5 mL tube. Centrifuge T cells 5 min at 1000g. Resuspend T cells in 50 μL of uncolored RPMI, HEPES 10 mM, 10% FBS.
2. Gently pipet 10 μL of T cell suspension with a pre-warmed tip then add it on a side to P815-immobilized Petri dish on a temperature-equilibrated block. Do not mix T cells with P815 and do not agitate the dish. Cover the cells with a glass slip avoiding bubbles if the temperature-controlled chamber of the microscope stage is not humidified to prevent drying.
3. Transfer carefully the dish from the temperature-equilibrated aluminum block to the temperature-controlled stage of the inverted transmission microscope and an oil immersion objective (40×, NA 1.3).
4. P815 cells can be differentiated from CEM cells by using fluorescence imaging mode. The CFSE Trace (ex 492 nm/em 517 nm) absorbed by P815 cells turn their strong fluorescence on. Focus on the adhered P815. Check the cell density. Look for a non-dense area where there is enough space all around a P815 to move or roll a CEM as shown in Fig. 3C.
5. Light on the near-infrared laser at low power for trapping a swimming T cell or a T cell sedimented on the glass floor with the joystick. One tweezer should be enough to manipulate one cell.

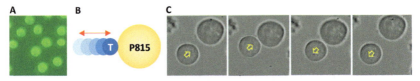

Fig. 3 Binding kinetics through repeated contact of T cell at the same spot on P815. (A) CFSE-labeled P815 to differentiate them from T cells. (B) Process. (C) Cell-cell contact. P815 is adhered to the glass floor, the T cell is moved by a single optical tweezer back and forth with a contact time of 0.2 s. Yellow arrows indicate the direction of the applied force.

6. Move the T cell close to the targeted P815, avoiding other cells, as any cells below or above it would be trapped. Actually, the joystick moves up-and-down and right-and-left the nano-stage holding the Petri dish with its bound cells while the trapped T cell remain immobile in the field of view.

3.4 Repetitive contacts of the same area of the two cells (Fig. 3)

1. Check for the stable adhesion of the selected P815 used for target. Put the optical tweezer on the P815 cells and try to move it up and down and left and right.
2. Place the cell at three times its diameter away from the P815 target at the same Z-level but not touching the glass floor.
3. Start the image recording.
4. Move the T cell in straight line up to contact with the P815 surface then move back to the initial position. Time of contact 0.2 s. Force feedback makes the contact easier to control.
5. Repeat the operation every second to the same spot at the surface of P815 until immobilization of the T cell.
6. Stop the image recording.
7. Count the number of contacts prior immobilization.

It is possible to automatize the repetitive motions by programming the displacement of the optical tweezer with a sinusoidal rate along the straight line between the starting and full contact position. The time of contact and time between contacts are more rigorous. If the optical tweezer is manually controlled use a beeper or a metronome (smartphone application) for cadencing contacts every seconds.

3.5 Repetitive contacts of the same area of the T cell and different spots around P815 (Fig. 4)

1. Select a P815 cell with open space around it. Check for its stable adhesion by jiggling the cell with the optical tweezer.

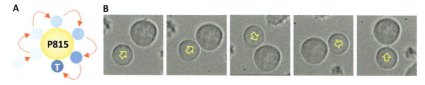

Fig. 4 Binding kinetics through repeated contact of T cell at different spots around P815. (A) Process. (B) Cell-cell contact. P815 is adhered to the glass floor, the T cell is moved by a single optical tweezer from spot to spot with a contact time of 0.2 s. Yellow arrows indicate the direction of the applied force.

2. Place the T cell at three times its diameter away from the P815 target at the same Z-level but not touching the floor.
3. Start the image recording.
4. Move the T cell in straight line up to contact with the P815 surface then move back to the initial position. Time of contact 0.2 s. Force feedback makes the contact easier to control.
5. Repeat the operation every second to a different spot around the surface of P815 until immobilization of the T cell.
6. Stop the image recording.
7. Count the number of contacts prior immobilization.

It is possible to automatize the repetitive motions by programming the displacement of the optical tweezer with a sinusoidal rate, bouncing around the target cell from one contact position to another, the time of contact is more rigorously monitored. If not use a beeper or a metronome.

3.6 Rolling T cell all around P815 until immobilization (Fig. 5)

1. Select a P815 cell with open space around it.
2. Place the cell at three times its diameter away from the P815 target at the same Z-level but not touching the floor.
3. Start the image recording.
4. Move the T cell in straight line up to contact with the P815 surface then roll it around the surface of P815 at constant speed with a constant force to ensure the continuous cell-cell contact until immobilization of the T cell.
5. Stop the image recording.

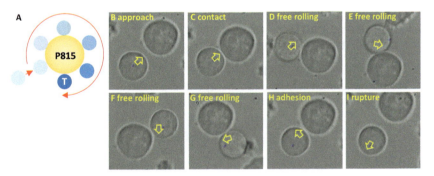

Fig. 5 Binding kinetics through rolling T cell all around P815. (A) Process, (B) approach, (C) cell-cell contact, (D-G) free rolling until (H) adhesion and (I) rupture by pulling away the T cell. P815 is adhered to the glass floor, the T cell is moved by a single optical tweezer tangentially to the cell-cell interface. Yellow arrows indicate the direction of the applied force.

6. Report the time and the rolling distance from contact to immobilization. It is possible to automatize the rolling motion by programming the displacement of the optical tweezer around the cell but we did not find any advantages.

3.7 Force data analysis

The force exerted by the optical tweezers on a cell is like that of a spring, proportional to the distance the cell is deflected from the position of the laser application point before and after the cell is resisted by contact with another cell or any other obstacle at an equilibrium position. The intensity of this force is calculated using the Hooke's law of spring model (Ashkin, 1992):

$$F_{OT} = K_{OT} \left(P_{laser} - P_{cell} \right)$$

where the laser position P_{laser} is fixed in the field of view reference, and P_{cell} represent the cell position as obtained from the tracking method. K_{OT} is the stiffness of the optical trap. The stiffness K_{OT} is calculated experimentally using the hydrodynamics approach (Chen et al., 2007). The optical tweezer was centered on the T cell and once trapped, the T cell was moved linearly to a constant speed while image series were recorded. The dragging force was calculated from Stokes' law:

$$F_{drag} = -6 \pi \eta R_s v_{cell}$$

where v_{cell} is the cell velocity calculated from image tracking, R_s is the Stokes radius of cell, and η is the dynamic viscosity of the solution at 37 °C. Since the dragging force equaled to the force of optical trap, the stiffness of optical trap K_{OT} could be calculated:

$$F_{OT} = -F_{drag} = K_{OT} \left(P_{laser} - P_{cell} \right) = 6 \pi \eta R_s v_{cell}$$
$$K_{OT} = 6 \pi \eta R_s v_{cell} / \left(P_{laser} - P_{cell} \right)$$

As given example, considering 300 mW laser power it is estimated in x-axis, y-axis and z-axis as:

$$K_{OT,x} = 11.8 \text{ pN}/\mu m, K_{OT,y} = 12.1 \text{ pN}/\mu m \text{ and } K_{OT,z} = 1.2 \text{ pN}/\mu m$$

respectively under temperature of 37 °C. Axial stiffness is 10 times less than lateral stiffness. This indicates that during manipulation, the loss of the trapped cell is more likely caused by the reacting force in the axial direction. The stiffness constant K_{OT} should be calibrated for any specific application when changing the experience conditions, i.e. cell size, temperature, laser power, medium, etc.

4. Results

The murine P815 mast cell lines used as a tumor model have been labeled with a fluorescent dye (CFSE) and then adhered to the glass bottom Petri dish (Fig. 3A). T cells were then added to the same Petri dish. Optical tweezers were used only to manipulate the T cells. The single optical tweezer is placed at the center of an isolated T cell and moved by sloloming between obstacles to bring it closer to well-attached target P815 cells with sufficient space around it. The T cell is placed in the same horizontal plane as the P815 cell, ensuring that it does not touch the glass floor. The T cell is brought into contact with the P815 cells to measure the contact time required at a single point, over several points or in the course of a rolling path to initiate adhesion under culture conditions.

When the T cell is moved linearly into contact with the P815 cell repeatedly at the same spot or at different spots every second for an average contact time of around 0.2 s, it takes more than 50 contacts before the T cell can no longer be released with a force of less than 100 pN. When P815 have been pre-incubated in the presence of anti-human CD3 antibodies (OKT3 0.1 μg/mL), 1–43 contacts (median 9.5) are required to initiate adhesion after contact at the same spot, 1–49 contacts (median 10) at different spots at the P815 surface (Fig. 6A). The cumulative times of contact along repeats until stable adhesion have been reported in Fig. 6B and corresponding

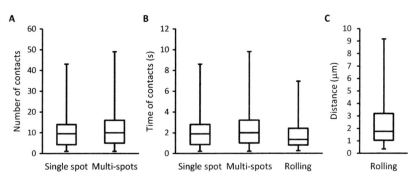

Fig. 6 (A) Number of repeated contacts required for adhesion on the same spot at the surface of OKT3-treated P815 (single spot; $n=50$) on different spots (multi-spots; $n=50$). The whisker-box plots indicate the quartiles: whiskers for min and max, boxes for second and third quartiles framing the median. (B) Sum of the contact times for single ($n=50$) and multi-spots ($n=50$) and time to adhesion when rolling the T cell on the OKT3-treated P815 surface ($n=50$). (C) Distance traveled by the T cell rolling the surface of the OKT3-treated P815.

distance Fig. 6C. These measurements have been repeated with 50 different T cell/P815 pairs for each condition.

Fig. 7 shows the T cell manually rolled with a constant force with the optical tweezer in a tight contact around the glass-bound P815 cell surface. The median time to immobilization by adhesion is more than 40 s and a path longer than 50 µm (Fig. 6A). After pre-incubation of T cells in the presence of anti-CD3 antibody (0.1 µg/mL), this median time is 1.3 s (from 0.2 s to 7 s) and the median distance traveled is 1.8 µM (Fig. 5C). Several examples are shown in the Fig. 6B-E with time to adhesion from 0.25 to 4.23 s. Fifty T cells have been rolled over different untreated and OKT3-treated P815 cells.

5. Concluding remarks

Measuring the adhesion speed of T cells to solid tumor cells provides crucial insights in documenting T cell surveillance and tumor cell recognition and how it controls tumor progression.

Here, we use CD3 engagement of the T cell to initiate binding to the P815 mast cell, by the anti-CD3 antibody OKT3 associated with its FcγRII receptor. OKT3 has a high affinity for CD3, and a strength of the order of several tens of pN. Ten sub-second contacts or a 1.3 s-continuous contact of T cells with the P815 cells are enough for initiating its adhesion, meaning resisting to 100 pN force, about the resistance against a local flow (Kamsma et al., 2018; Lehenkari & Horton, 1999). The same number of contacts is necessary to initiate the adhesion on the same or different spots at the P815 surface. This observation suggests that T cell activation of the adhesion process is independent of an activation process on the P815 side. The cumulative time of the repeated contact are about 2 s, the same duration time of T cell rolling at the P815 surface (median 1.3 s). This suggests that the adhesion initiation is integrating the contact steps and there is either (1) la propagation of this activation process as repetitive contacts engaged the very same spot, different spots or all along the traveled surface of P815 when T cell is rolling, or (2) the probability of encountering CD3 on the T cell and anti-CD3 on P815 is a function of time during the contact period. Of course, the time and path length of T cells prior adhesion to P815 cells depends on rolling velocity, challenging the occurrence of bonds. We tried in this study to keep the T cell speed constant as much as possible on P815 surface (1.3 µm/s), about 20 s to roll a T cell around a P815. Haptic system with force feedback makes the control of applied force easier when rolling cells (Gerena et al., 2020, 2023). However, in the absence of feedback as in most of commercial equipment settings, visual

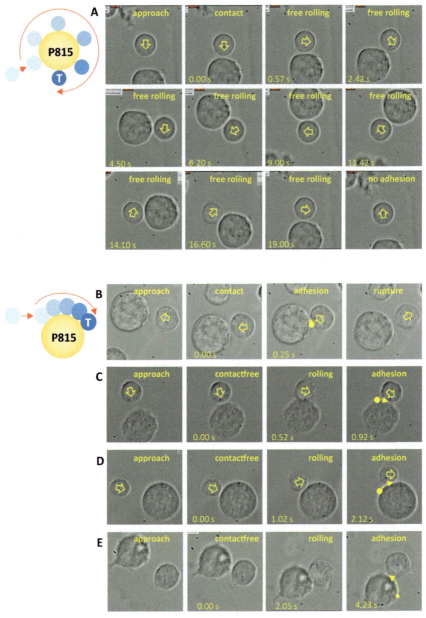

Fig. 7 Binding kinetic assays of T cell on P815 cells in the absence (A) and the presence (B-E) of OKT3 (0.1 μg/mL) by rolling the T cell with the optical tweezer over the glass surface-bound P815 cells. Successive images from approach, contact, rolling to adhesion are presented with the corresponding times since contact. Tweezer force direction is indicated with an arrow. Laser power is increased to break the interface and separate cells, when possible, as shown in (B). If the maximum force of 500 pN applied by the optical tweezers to the cell is not enough to break the interaction, as in examples C to E, the cells will remain attached to each other.

monitoring of cell rotation and membrane deformation is sufficient to control the contact force at the cell interface after some training.

Immune cell adhesion kinetics to tumor cells could potentially be use as severity assay of the lack of immune response, illustrating the escape of T cells or the recruitment of T cell promoting tumor growth. T cell adhesion speed assays can be used to screen and identify molecules responsible for cell adhesion and potential modulators that could be of therapeutic interests for reducing the early recruitment of T lymphocytes in the tumor microenvironment in order to slow down tumor growth.

Accordingly, thus method of measuring the adhesion speed of immune cells to tumor cells offers significant tools to study adhesion, synapse dynamics to understand tumor biology, and application for prognosis, diagnosis and therapy follow-up.

6. Perspectives

The system upgrading with fluorescence imaging capabilities allows to visualize using fluorescent probes, the cytoskeleton organization of the immune and the target cell in the course of their interaction leading to the formation of the synapse. Several projects involving measurements of the interaction kinetics of immune cells with different solid tumor cell targets are also running, focused on CD4 and CD8 T lymphocytes and natural killers in order to study the structural and functional construction of immune synapses.

7. Notes

1. If the target cells are not well adhered, it is possible to increase poly-lysine concentration to reinforce the adhesion or decrease the concentration of FBS during adhesion process. Neutralization of poly-lysine with 10% FBS (or at least 2 mg/mL of BSA) is necessary then before adding the T cell to move.

2. If the target cells are adherent, they should be placed in culture conditions for several hours or even overnight to promote their spreading and attachment, renewing the warmed medium (37 °C) requires to pre-equilibrate it with dissolved gas (5% CO_2) in a bottle featuring a gas-permeable cap.

3. The density of adherent cells should be low with open space around it to drive mobile cells around.

4. Target cells could be surface-adhered from flat to spherical shapes but it is easier with thick cells even if not spherical. Similar investigations can be performed with single cells, cell patches or organoids.

Globally these experiments require organization and long training with the 7D-joystick. To obtain accurate data, there are a number of delicate points to consider.

1. Cells should never be allowed to fall below 32 °C in the 30 min prior to manipulation, to avoid affecting adhesion kinetics and strength.
2. Cells should not be manipulated more than a minute with the near-infrared laser since it will warm up the cell and could produce toxic free radicals (Blázquez-Castro, 2019). Take time to choose the cell to move and the target to avoid long manipulation with laser exposure. Switch off the laser beam as soon as possible.
3. Prepare cells in several Petri dish with the same conditions. Change Petri dish every half hour.
4. Manipulating 50 T cells for each of the three protocols takes about 2 h. It is necessary to use several dishes for completing the assays.

Several facts may affect data analysis.

1. The moving cell pressure to the adherent cells should be roughly constant, force feedback helps. The pressure should be enough to see the moving cell rolling not sliding.
2. Rolling speed affects the time and path length of T cells prior adhesion to OKT3-treated P815, challenging the formation of bonds. A constant velocity should be maintained around the cell.

Acknowledgments

The authors thank Dr. Yves Janin for his comments and proofreading. This work was supported by grants from La Ligue Nationale contre le Cancer (Equipe Labellisée Ligue 2018) from the Agence Nationale de la Recherche (ANR 2021 CE33 0003 03 OPTOBOT), region Ile-de-France (DIM Elicit equipment) and institutional grants from Institut Pasteur. The Diagnostic Test Innovation & Development core facility is supported by the grant SESAME from the French public investment bank (BPI), region Ile-de-France and Institut Pasteur. EG has been funded by the Tech Transfer agency of Sorbonne Université and NI has been funded by the ANR grant.

Declaration of interest

The authors declare that they have no conflict of interest.

References

Aguera-Gonzalez, S., Burton, O. T., Vazquez-Chavez, E., Cuche, C., Herit, F., Bouchet, J., et al. (2017). Adenomatous polyposis coli defines Treg differentiation and anti-inflammatory function through microtubule-mediated NFAT localization. *Cell Reports, 21,* 181–194. https://doi.org/10.1016/j.celrep.2017.09.020.

Ashkin, A. (1992). Forces of a single-beam gradient laser trap on a dielectric sphere in the ray optics regime. *Biophysical Journal, 61*(2), 569. https://doi.org/10.1016/S0006-3495(92) 81860-X.

Ashkin, A., Dziedzic, J., & Yamane, T. (1987). Optical trapping and manipulation of single cells using infrared laser beams. *Nature*, *330*, 769–771. https://doi.org/10.1038/330769a0.

Benoit, M., Gabriel, D., Gerisch, G., & Gaub, H. E. (2000). Discrete interactions in cell adhesion measured by single-molecule force spectroscopy. *Nature Cell Biology*, *2*, 313–317. https://doi.org/10.1038/35014000.

Benoliel, A., Capo, C., Mege, J. L., & Bongrand, P. (1994). Measurement of the strength of cell-cell and cell-substratum adhesion with simple methods. In *Studying cell adhesion* (pp. 81–92). Springer-Verlag.

Blázquez-Castro, A. (2019). Optical tweezers: Phototoxicity and thermal stress in cells and biomolecules. *Micromachines (Basel)*, *10*(8), 507. https://doi.org/10.3390/mi10080507.

Chen, H. D., Ge, K. K., Li, Y. M., Wu, J. G., Gu, Y. Q., Wei, H. M., et al. (2007). Application of optical tweezers in the research of molecular interaction between lymphocyte function associated antigen-1 and its monoclonal antibody. *Cellular & Molecular Immunology*, *4*(3), 221–225.

Favre-Bulle, I. A., & Scott, E. K. (2022). Optical tweezers across scales in cell biology. *Trends in Cell Biology*, *32*(11), 932–946. https://doi.org/10.1016/j.tcb.2022.05.001.

Foley, G. E., Lazarus, H., Farber, S., Uzman, B. G., Boone, B. A., & McCarthy, R. E. (1965). Continuous culture of human lymphoblasts from peripheral blood of a child with acute leukemia. *Cancer*, *18*, 522–529. https://doi.org/10.1002/1097-0142(196504)18:4<522::aid-cncr2820180418>3.0.co;2-j.

Gajewski, T. F., Markiewicz, M. A., & Uyttenhove, C. (2001). The p815 mastocytoma tumor model. *Current Protocols in Immunology*, *Chapter 20*. https://doi.org/10.1002/0471142735.im2004s43. Unit 20.4.

Gerena, E., & Haliyo, S. (2023). 3D force-feedback optical tweezers for experimental biology. In C. Dai, G. Shan, & Y. Sun (Eds.), *Robotics for cell manipulation and characterization* (pp. 145–172). Academic Press. https://doi.org/10.1016/B978-0-323-95213-2.00010-7. 9780323952132.

Gerena, E., Legendre, F., Molawade, A., Vitry, Y., Régnier, S., & Haliyo, S. (2019). Tele-robotic platform for dexterous optical single-cell manipulation. *Micromachines*, *10*(10), 677. https://doi.org/10.3390/mi10100677.

Gerena, E., Legendre, F., Vitry, Y., Regnier, S., & Haliyo, S. (2020). Improving optical micromanipulation with force-feedback bilateral coupling. In *2020 IEEE international conference on robotics and automation (ICRA)* (pp. 10292–10298). IEEE. https://doi.org/10.1109/ICRA40945.2020.9197424.

Gosse, C., & Croquette, V. (2002). Magnetic tweezers: Micromanipulation and force measurement at the molecular level. *Biophysical Journal*, *82*, 3314–3329. https://doi.org/10.1016/S0006-3495(02)75672-5.

Hogan, B., Babataheri, A., Hwang, Y., Barakat, A. I., & Husson, J. (2015). Characterizing cell adhesion by using micropipette aspiration. *Biophysical Journal*, *109*(2), 209–219. https://doi.org/10.1016/j.bpj.2015.06.015.

Jahnel, M., Behrndt, M., Jannasch, A., Schäffer, E., & Grill, S. W. (2011). Measuring the complete force field of an optical trap. *Optics Letters*, *36*(7), 1260–1262. https://doi.org/10.1364/OL.36.001260. PMID: 21479051.

Juzans, M., Cuche, C., Rose, T., Mastrogiovanni, M., Bochet, P., Di Bartolo, V., et al. (2020). Adenomatous polyposis coli modulates actin and microtubule cytoskeleton at the immunological synapse to tune CTL functions. *Immunohorizons*, *4*, 363–381. https://doi.org/10.4049/immunohorizons.2000044.

Kamsma, D., Bochet, P., Oswald, F., Ablas, N., Goyard, S., Wuite, G. J. L., et al. (2018). Single-cell acoustic force spectroscopy (scAFS): Resolving kinetics and strength of T-cell adhesion to fibronectin. *Cell Reports*, *2018*(24), 3008–3016. https://doi.org/10.1016/j.celrep.2018.08.034.

Landegren, U., Andersson, J., & Wigzell, H. (1984). Mechanism of T lymphocyte activation by OKT3 antibodies. A general model for T cell induction. *European Journal of Immunology*, *14*(4), 325–328. https://doi.org/10.1002/eji.1830140409.

Lasserre, R., Charrin, S., Cuche, C., Danckaert, A., Thoulouze, M. I., de Chaumont, F., et al. (2010). Ezrin tunes T-cell activation by controlling Dlg1 and microtubule positioning at the immunological synapse. *The EMBO Journal, 29*, 2301–2314. https://doi.org/10.1038/emboj.2010.127.

Lehenkari, P. P., & Horton, M. A. (1999). Single integrin molecule adhesion forces in intact cells measured by atomic force microscopy. *Biochemical and Biophysical Research Communications, 259*, 645–650. https://doi.org/10.1006/bbrc.1999.0827.

Li, F., Redick, S. D., Erickson, H. P., & Moy, V. T. (2003). Force measurements of the alpha5beta1 integrin-fibronectin interaction. *Biophysical Journal, 84*, 1252–1262. https://doi.org/10.1016/S0006-3495(03)74940-6.

Mastrogiovanni, M., Di Bartolo, V., & Alcover, A. (2022). Cell polarity regulators, multifunctional organizers of lymphocyte activation and function. *Biomedical Journal, 45*(2), 299–309. https://doi.org/10.1016/j.bj.2021.10.002.

Mastrogiovanni, M., Juzans, M., Alcover, A., & Di Bartolo, V. (2020). Coordinating cytoskeleton and molecular traffic in T cell migration, activation, and effector functions. *Frontiers in Cell and Development Biology, 8*, 591348. https://doi.org/10.3389/fcell.2020.591348.

Nieminen, T. A., Knöner, G., Heckenberg, N. R., & Rubinsztein-Dunlop, H. (2007). Physics of optical tweezers. *Methods in Cell Biology, 82*, 207–236. https://doi.org/10.1016/S0091-679X(06)82006-6.

Norman, D. J. (1995). Mechanisms of action and overview of OKT3. *Therapeutic Drug Monitoring, 17*(6), 615–620. https://doi.org/10.1097/00007691-199512000-00012.

Schaffer, E., Norrelykke, S. F., & Howard, J. (2007). Surface forces and drag coefficients of microspheres near a plane surface measured with optical tweezers. *Langmuir, 23*, 3654–3665. https://doi.org/10.1021/la0622368.

Schindelin, J., Arganda-Carreras, I., Frise, E., Kaynig, V., Longair, M., Pietzsch, T., et al. (2012). Fiji: An open-source platform for biological-image analysis. *Nature Methods, 9*(7), 676–682. https://doi.org/10.1038/nmeth.2019.

Schneider, C. A., Rasband, W. S., & Eliceiri, K. W. (2012). NIH Image to ImageJ: 25 years of image analysis. *Nature Methods, 9*(7), 671–675. https://doi.org/10.1038/nmeth.2089.

Shao, J. Y., & Hochmuth, R. M. (1996). Micropipette suction for measuring piconewton forces of adhesion and tether formation from neutrophil membranes. *Biophysical Journal, 71*(5), 2892–2901. https://doi.org/10.1016/S0006-3495(96)79486-9.

Simpson, K. H., Bowden, M. G., Höök, M., & Anvari, B. (2002). Measurement of adhesive forces between S. epidermidis and fibronectin-coated surfaces using optical tweezers. *Lasers in Surgery and Medicine, 31*(1), 45–52. https://doi.org/10.1002/lsm.10070.

Sitters, G., Kamsma, D., Thalhammer, G., Ritsch-Marte, M., Peterman, E. J., & Wuite, G. J. (2015). Acoustic force spectroscopy. *Nature Methods, 12*, 47–50. https://doi.org/10.1038/nmeth.3183.

van Mameren, J., Wuite, G. J., & Heller, I. (2011). Introduction to optical tweezers: Background, system designs, and commercial solutions. *Methods in Molecular Biology, 783*, 1–20. https://doi.org/10.1007/978-1-61779-282-3_1.

Wang, X., Chen, S., Kong, M., Wang, Z., Costa, K. D., & LI, R.A., Sun, D. (2011). Enhanced cell sorting and manipulation with combined optical tweezer and microfluidic chip technologies. *Lab on a Chip*, 3656–3662. https://doi.org/10.1039/C1LC20653B.

Yin, M., Gerena, E., Pacoret, C., Haliyo, S., & Régnier, S. (2017). High-bandwidth 3D force feedback optical tweezers for interactive bio-manipulation. In *2017 IEEE/RSJ international conference on intelligent robots and systems (IROS)* (pp. 1889–1894). IEEE. https://doi.org/10.1109/IROS.2017.8206006.

CHAPTER EIGHT

Measuring interaction force between T lymphocytes and their target cells using live microscopy and laminar shear flow chambers

Sophie Goyard[a], Amandine Schneider[a,b], Jerko Ljubetic[a,b], Nicolas Inacio[a,c], Marie Juzans[b], Céline Cuche[b], Pascal Bochet[d], Vincenzo Di Bartolo[b], Andrés Alcover[b], and Thierry Rose[a,*]

[a]Institut Pasteur, Université Paris Cité, Diagnostic Test Innovation & Development Core Facility, Paris, France
[b]Institut Pasteur, Université Paris Cité, INSERM-U1224, Unité Biologie Cellulaire des Lymphocytes, Ligue Nationale Contre le Cancer, Équipe Labellisée Ligue-2018, Paris, France
[c]Institut des Systèmes Intelligents et de Robotique (ISIR), Sorbonne Université, CNRS, Paris, France
[d]Institut Pasteur, Université de Paris Cité, Image Analysis Unit, CNRS UMR3691, Paris, France
*Corresponding author: e-mail address: thierry.rose@pasteur.fr

Contents

1.	Introduction	176
	1.1 T-cell adhesion target cell adhesion process	176
	1.2 Mechanobiology toolbox for measuring rupture force of immune cell pair interactions	177
2.	Material	179
	2.1 Disposables	179
	2.2 Reagents	179
	2.3 Cells	180
	2.4 Equipment	180
	2.5 Software	181
3.	Methods	181
	3.1 Cell culture (2 days)	181
	3.2 CD8+ T cell lentiviral infection to generate APC-silenced CTLs (3 days)	182
	3.3 CD8+ T cells for control (60 min)	182
	3.4 Slide preparation (90 min)	183
	3.5 Target cell immobilization (60 min)	184
	3.6 Slide installation, immune cell addition (15 min)	185
	3.7 Flow rate calibration (5 min)	185
	3.8 Flow rate gradient and image acquisition (5 min)	186
	3.9 Image analysis	187
	3.10 Data analysis	188
4.	Results	192
5.	Concluding remarks	194

Methods in Cell Biology, Volume 193
ISSN 0091-679X
https://doi.org/10.1016/bs.mcb.2024.09.001

6. Perspectives	194
Notes	195
Ethics statement	197
Acknowledgments	197
Declaration of interest	198
References	198

Abstract

Understanding the immunological synapse formation and dynamics can be enriched by measuring cell-cell interaction forces and their kinetics. Microscopy imaging reveals structural organization of the synapse, while physical methods detail its mechanical construction. Various techniques have been reported for measuring forces needed to rupture the interface between a T lymphocyte and its target cell but most of them measure one pair at a time. We describe here a laminar shear flow-based method that exerts dragging forces on T cell-target cells pairs immobilized on the surface of a flow chamber. Increasing flow rate allows us to observe the detachment of hundreds of cell conjugates on the wide field of a light transmission microscope. Monitoring precisely the flow rate gradient exerted on T cells readily yields synapse rupture measurements. Dragging forces are measured at the point of rupture as a linear function of the flow speed in minutes from 10 pN to 20 nN for each cell pair among a statistically representative cell population in the whole field of view of a single experiment. The output cells can be collected in multi-well plate sorted in the increasing order of rupture forces. We used this approach to unveil the involvement of the cytoskeleton regulator adenomatous polyposis coli (APC) in the stability of immunological synapses formed between human cytotoxic T cell and tumor target cells. APC is a polarity regulator and tumor suppressor associated with familial adenomatous polyposis and colorectal cancer. Reduced APC expression impairs T cell adhesion with tumor target cells suggesting an impact of APC mutation in anti-tumor immune defense.

1. Introduction
1.1 T-cell adhesion target cell adhesion process

The formation of immunological synapses between T cells and their target cells is a complex mechanism whose initial steps are based on a process of contact, signal transduction and then activation leading to the progressive reinforcement of the interaction of the T cell with the target cell via the structural and functional reorganization of T cell integrin molecules, LFA-1 and VLA-4 with their cognate ligands on the target cells, ICAM-1 and VCAM-1. Integrins expressed on T cells are inactive in their resting state. A process of inside-out signaling driven by the T cell antigen receptor (TCR) lead to integrin conformation changes, binding to cortical cytoskeleton and clustering and, as a consequence enhanced affinity and avidity for their ligands. In addition, outside-in integrin-mediated signaling also reinforces their adhesion capacity (Dustin & Springer, 1989; Kinashi, 2012).

1.2 Mechanobiology toolbox for measuring rupture force of immune cell pair interactions

The cell-cell adhesive strength is measured by determining the force required to rupture their contact. Several physical methods have been developed to pull apart cell pairs and thus measure the force required forces. They can be divided into two groups. In a first group, atomic force microscopy (AFM) (Benoit, Gabriel, Gerisch, & Gaub, 2000; Lehenkari & Horton, 1999; Li, Redick, Erickson, & Moy, 2003), micropipette aspiration (MA) (Hogan, Babataheri, Hwang, Barakat, & Husson, 2015; Shao & Hochmuth, 1996), optical tweezer (OT; Ashkin, Dziedzic, & Yamane, 1987; Schaffer, Norrelykke, & Howard, 2007; Wang et al., 2011; Favre-Bulle & Scott, 2022) or magnetic tweezers (MT; Gosse & Croquette, 2002) exert forces on one pair of cells at a time. In a second group, laminar shear flow devices (LSF) (Benoliel, Capo, & Mege and P. Bongrand., 1994; Lu et al., 2004), acoustic force spectroscopy (AFS) (Kamsma, Creyghton, Sitters, Wuite, & Peterman, 2016; Kamsma et al., 2018; Sitters et al., 2015), shear-spinning disks (SSD) (Boettiger, 2007; Friedland, Lee, & Boettiger, 2009; Garcia, Huber, & Boettiger, 1998) and centrifugation (Reyes & Garcia, 2003) exert forces on multiple pairs simultaneously. Light microscopy events monitoring is used either at start and end points or in real time as continuous series in all methods. Surface plasmon resonance has been also used as alternative to imaging for measuring cell association and dissociation in the course of adhesion process (Kim, Chegal, Doh, Cho, & Moon, 2011). The type of forces exerted by AFM, MA, OT, OM, AFS and centrifugation are orthogonal to the interface plane, or exerted oppositely on both sides of the interface, ruptures starting from the periphery to the center of the interface. On the other hand, the forces exerted by LSF and SSD are lateral but mainly tangential as the flow is exerted asymmetrically, from high on cell top to low on cell bottom and provokes a rotation of the cell. Therefore, the force is not exerted with the same intensity across the interface and integrin-ligand pair interaction ruptures occur sequentially as a wave from front to rear with orthogonal traction on each pair.

The rupture force also depends on the loading rate (Merkel, Nassoy, Leung, et al., 1999). The rupture is a continuous reorganization of molecular structure where the increasing distance between integrins and their target proteins is making the process irreversible. The high loading rate challenges the rearrangement of broken pairs.

Some forces are exerted by contact with a rigid object which can itself induce activation and structuring of the cytoskeleton such as the AFM stylus, or the MA micropipette. Moreover, optical tweezer can induce localized photoactivation and formation of toxic free radicals. Other methods do not exert any rigid physical contact. AFS exerts acoustic plane wave pressure generated by a piezoelectric crystal through the medium on any density contrast between medium and cell. Centrifugation exerts gravitational forces on cell. As in the case of the SSD, the LSF has the advantage of using the dragging force generated by the flow rate of the culture medium without further contact to detach the immune cells from the target cells immobilized on the chamber floor. The SSD allows the measurement of the forces according to the distance from the center of the friction disk which rotates above the floor disk where the target cells are fixed.

Concerning the laminar shear flow, its principle is very simple and the method has been used and developed for years (Kaplanski et al., 1993; Olivier & Truskey, 1993; Benoliel et al., 1994; Pierres, Tissot, & Bongrand, 1994, Pierres, Tissot, Malissen, & Bongrand, 1994; Chapman & Cokelet, 1996; Robert, Limozin, van der Merwe, & Bongrand, 2021). The rupture force of the interaction of the lymphocyte with its target is measured in an increasing flow rate which exerts a dragging force ramp. It is also possible to observe 10 of pairs on the microscope field during a single experiment and measure the rupture force for each single pair in the same experiment, under the same conditions. Thus, this simple and robust method allows to exert in seconds or minutes, synapse rupture force values ranging from 10 pN to 20 nN using a monitored increasing flow rate providing force ramp between 1 and 500 pN/s on a statistically representative sample of a cell population observed in the field of view.

In the present work, mouse mastocytes P815 cell line, a mastocyma tumor cell model, were immobilized in the polylysin–coated flow chamber between two glass slides, then treated with anti-CD3 antibodies, then human activated CD8 lymphocytes were injected under culture conditions while the CD8-P815 pair forming were monitored by light transmission microscopy. Then by using computer-controlled syringe pump, an increasing flow rate of cell culture medium at 37 °C passes from one end through the thermostated chamber and is discarded to a waste bottle. Alternatively, cells may be collected from this end into a multi-well plate fraction collector to recover the released cells fractionated according to the flow rate for further analysis or culture expansion. The dragging force necessary to break the interaction were compared using control and patients CD8 T lymphocytes or under conditions of silencing of the cytoskeleton modulator APC in CD8 T lymphocyte to identify APC involvement in the synapse formation and its mechanical function (Juzans et al., 2020; Juzans, Cuche, Di Bartolo, & Alcover, 2023).

2. Material

2.1 Disposables

1. 10 μL, 200 μL, 1000 μL tips (Sorenson)
2. 24-well plates for cell culture (Falcon, Ref# 353047).
3. Sticky μ-Slide I$^{0.1}$ Luer (Ibidi, Ref# 81128).
4. Sterile microcentrifuge tubes (1.5 mL, Eppendorf Ref# 3810).
5. 96-well plates for cell culture (TPP, Ref# 92097).
6. Glass slide 70 × 25 × 1 mm (Rogo-Sampaic)
7. Petri plate for cell culture (TPP, Sigma-Aldrich, Ref# Z707678-840EA)
8. Flask 25 cm^2 (Clearline, Dominique Dutcher, Ref# 131000C)
9. LS columns for magnetic cell sorting (Miltenyi Biotec, Ref# 130-042-401).
10. 2 mL syringe (BD) with concentric Luer slip for slide loading
11. 60 mL syringe (Omnifix Braun) with excentred Luer slip used with the syringe pump
12. Recycled PBS bottles (500 mL)

2.2 Reagents

1. Sulfuric acid (Sigma-Aldrich, Ref# 258105)
2. Sodium hydroxide (SAFC, Ref# 1.37023.1002)
3. Hydrogen peroxide 33% (Sigma-Aldrich, Ref# H1009)
4. Phosphate buffered saline (PBS, Gibco, ThermoFisher Scientific, Ref# 14190).
5. Bovine fibronectin (Sigma-Aldrich, Ref# F4759)
6. ICAM-1 (R&D, 720-IC-200) [for CTL adhesion force assay on ICAM-1-functionalized glass]
7. rhVCAM-1/Fc Chimera (R&D Systems, 862-VC [for CTL adhesion force assay on VCAM-1-functionalized glass]
8. Bovine serum albumin (BSA, Alpha Diagnostics, Ref# 80400-100).
9. Ethanol (Sigma-Aldrich, Ref# 100983)

10. Roswell-Park Memorial Institute medium (RPMI) 1640 medium with GlutaMAX-I and Phenol Red (Gibco, ThermoFisher Scientific, Ref# 61870).
11. RPMI 1640 medium, without phenol red (Gibco, ThermoFisher Scientific, Ref# 11835).
12. Sodium pyruvate (Life Technologies, Ref# 11360).
13. Nonessential amino acids (Life Technologies, Ref# 11140).
14. HEPES (Life Technologies, Ref# 15630-056).
15. Penicillin–streptomycin (Gibco, Ref# 15140-122).
16. Human serum (Dominique Dutscher, Ref# S4190-100).
17. Fetal bovine serum (FBS, Dominique Dutscher, Ref# SV30160.03).
18. Anti-CD3 antibody (Invitrogen, clone: OKT3, 16-0037-85)
19. Lymphocyte Separating Medium Pancoll Human tubes (Pan Biotech, Ref# P04-60125).
20. Magnetic cell sorting CD8+ T cell isolation kit (Miltenyi Biotec, Ref# 130-096-495).
21. Purified anti-human CD3ε antibody, clone UCHT1 (BioLegend Inc., Ref# 300402).
22. Anti-human CD28 antibody (Beckman Coulter, Ref# IM1376).
23. Recombinant human IL-2 (PeproTech, Ref# 200-02).
24. Lentiviruses expressing shRNAs (Sigma-Aldrich)
25. Puromycin (Gibco, Ref# A11138-03).
26. Cytochalasine-D (Sigma-Aldrich, Ref# C2618)
27. Colchicine (Sigma-Aldrich, Ref# C3915)

2.3 Cells

1. Mouse mastocyte P815 cell line (TIB-64, ATCC)
2. Human peripheral blood T cells from healthy volunteers from the French Blood Bank Organization (Etablissement Français du Sang) and from patients from the Association Polyposes Familiales France and the Institut Pasteur ICAReB Biological Resources core facility using ethically approved procedures.

2.4 Equipment

1. Cell culture incubator allowing standard cell culture conditions in a humidified atmosphere (37 °C, 5% CO_2)
2. Standard bench top centrifuge (Eppendorf, Ref# 5810R or 5415R)
3. Centrifuge for multi-well plates (Sigma, Ref# 4-15C)
4. Sterile cell culture laminar flow hood safety level II

5. MACS MultiStand (Miltenyi Biotec, Ref# 130-042-303)
6. MidiMACS™ Separator (Miltenyi Biotec, Ref# 130-042-302)
7. Light transmission inverted microscope (Axio Observer D1, Zeiss)
   a. objective 10×/0.3 numerical aperture (Zeiss)
   b. camera (Hamamatsu, ORCA-R)
   c. motorized stage (Applied Scientific Instrument, MS-2000)
   d. temperature-controlled chamber (Zeiss, TC 37_2)
8. Computer-monitored syringe pump (World Precision Instruments, SP210iW)
9. ETFE tubing (3M, Cytiva, Ref# 18114238)
10. Computer-controlled fraction collector in multi-well plate (Gilson, FC303B)
11. Monocanal pipettes (ThermoFisher, Finnpipettte F1)
12. Thermostating pouch 200×200 mm

2.5 Software

1. ImageJ (Schneider, Rasband, & Eliceiri, 2012) or Fiji (Schindelin et al., 2012) for image analysis.
2. Excel (Microsoft) or Prism (GraphPad) for data analysis.
3. The inlab-coded program in Python was developed for driving the image acquisition, the syringe flow rate and the fraction collector. It uses Micro-Manager (Edelstein et al., 2014) which propose a large panel of drivers.

3. Methods
3.1 Cell culture (2 days)

1. P815 cell line was grown in DMEM supplemented with 10% FBS, 10 mM HEPES, 1% penicillin–streptomycin (v/v) incubated in flask at 37 °C with 5% CO_2 in an incubator with humidified atmosphere.
2. PBMCs were isolated from healthy donor blood samples by density gradient centrifugation using Lymphocyte Separating Medium Pancoll Human.
3. $CD8^+$ T cells were isolated from PBMCs using the MACS $CD8^+$ T Cell Isolation Kit and maintained in human $CD8^+$ medium: RPMI 1640 plus GlutaMAX-I supplemented with 10% FBS, 1 mM sodium pyruvate and nonessential amino acids, 10 mM HEPES, 1% penicillin–streptomycin (v/v).
4. CTLs were generated from freshly isolated $CD8^+$ T cells stimulated for 2 days in 24 well plate with coated anti-CD3 (10 μg/mL; UCHT1),

soluble anti-CD28 (7 µg/mL), and from recombinant human IL-2 (100 U/mL) in 1 mL human CD8 medium per well.

3.2 CD8$^+$ T cell lentiviral infection to generate APC-silenced CTLs (3 days)

1. After 2-day activation of CTL in 24-well plates, centrifuge plates for 7 min at 450 × g and remove supernatant, add 900 µL of culture medium in which FBS is replaced by 10% human serum, supplemented with 100 U/mL IL-2, add 100 µL of lentiviruses coding for either control or *apc*-specific short hairpin RNAs (shRNAs) (10% v/v). Incubate at 37 °C with 5% CO_2 for 24 h.

2. Wash cells 3 times by centrifuging plates for 7 min at 450 × g, remove supernatant, add 1 mL of fresh culture medium. Repeat twice with medium supplemented with 100 U/mL IL-2 and 3.9 µg/mL puromycin. Cells can be taken out of the P2 laboratory. Incubate at 37 °C with 5% CO_2 for 3 days for selection. The percentage of infected CTLs was 70–85%, assessed by GFP expression by FACS.

3. The night before use, wash cells to remove dead cells. Centrifuge plates for 7 min at 450 × g, remove supernatant, add 1 mL of fresh culture medium with 100 U/mL IL-2 without puromycin.

4. Centrifuge the cells 5 min at 450 × g and resuspend in 100 µL RPMI HEPES 10 mM. Ready to use, keep at 37 °C.

3.3 CD8$^+$ T cells for control (60 min)

1. After 2-day activation of CTL from healthy donor and a patient suffering of adenomatous polyposis with qualified allelic altered *apc* gene in 24-well plates, the plate was centrifugated for 7 min at 450 × g, the supernatant removed, add 900 µL of culture medium in which FBS is replaced by 10% human serum, supplemented with 100 U/mL IL-2 and incubated overnight at 37 °C.

2. Non-infected CTL from healthy donor were resuspended in RPMI HEPES 10 mM in three wells: one for control, one with Cytochalasin D (10 µM final), one with Colchicine (10 µM final). Non-infected CTL from patient were resuspended in RPMI HEPES 10 mM. Incubate cells for 30 min at 37 °C.

3. Centrifuge the cells 5 min at 450 × g and resuspend in 100 µL RPMI HEPES 10 mM. Ready to use in the next 10 min, keep at 37 °C.

3.4 Slide preparation (90 min)

1. Take a standard glass slide (70 × 25 × 1mm) and clean and degrease its surface with ethanol. Remove any dust. Dust will cause leaks.
2. Remove the adhesive protection off the μ-Slide.
3. Assemble the clean standard glass slide to the μ-Slide to build the flow chamber (volume 25 μL) (Fig. 1). Press strongly with the thumbs all around the μ-Slide on top of the glass slide on the bench.
4. Fill a 2 mL-syringe with deionized water and place it to one Luer end. Fill the chamber and the Luer at the other end by pushing the syringe. Put a finger on top of the Luer and push the syringe for pressurizing the chamber and watch for leaks. Empty the water.
5. Clean by loading 100 μL (4 chamber volumes, CV = 25 μL) of sulfuric acid 2 M with a pipette, in Luer on one end by aspiring using the syringe at the other end, keep the channel full and incubate 5 min.
6. Activate the glass surface using 4 CV of hydrogen peroxide (33%), incubate 5 min.
7. Empty and rinse the syringe with deionized water and fill it with 2 mL of water.
8. Rinse the Luer and chamber with 2 mL of deionized water.
9. Empty and rinse the syringe with deionized water and fill it with 2 mL of water.

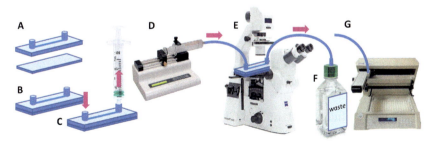

Fig. 1 Assembly of the laminar flow chamber. (A,B) Assembly of the μ-Slide with the glass slide. (C) Assembly of the syringe to one Luer and load to the opposite Luer for introducing the solution into the chamber (50 × 5 × 0.1 mm, volume 25 μL). When coated with target cells and loaded with CTL, the μ-Slide is posted on the thermostat-controlled stage of the inverted microscope, connected on one side to the syringe pump (D) and on the other side to the waste bottle (F) or a fraction collector (G).

184   Sophie Goyard et al.

10. Connect the syringe on the other Luer and rinse the Luer and chamber with 2 mL of deionized water.
11. Equilibrate with 12 CV of PBS, incubate 5 min.
12. Functionalize with 2 CV of bovine fibronectin (10 µg/mL in PBS) avoid air bubbles, incubated 1 h at room temperature.
13. Rinsed with 12 CV of PBS.

Assembled µ-Slides are reusable 30 to 50 times if they are properly cleaned after use with 10 CV of sodium hydroxide 2 M incubated 15 min, rinsed with 2 mL of water then air dried.

3.5 Target cell immobilization (60 min)

1. Prewarm the microscope stage 37 °C 1 h.
2. Fill the Luer with 100 µL of DMEM 10 mM (for avoiding bubbles) remove the syringe at the other end.
2. Fill the two Luers to their top with DMEM then put the syringe back.
3. Place the µ-Slide on aluminum block of dry bath at 37 °C and cover the chamber with a plastic lid for 10 min.
4. Empty the loading Luer by aspiring the medium using the syringe. From this time and on the chamber should never be empty until the end of the shear flow experiment.
5. Load with 37 °C-prewarm tips 80 µL of P815 cells suspended in DMEM HEPES 10 mM.
6. Empty the loading Luer by aspiring slowly the medium using the syringe keeping the chamber full.
7. Fill the Luer with 100 µL of DMEM avoiding evaporation in the chamber.
8. Cover the slide with the plastic lid, incubate 30 min at 37 °C.
9. Empty the loading Luer by aspiring slowly the medium using the syringe.
10. Load carefully 50 µL of anti-CD3 OKT3 (20 µg/mL final) in DMEM HEPES 10 mM.
11. Empty the loading Luer by aspiring slowly the medium using the syringe keeping the chamber full with the anti-CD3 OKT3 medium solution.
12. Fill the Luer with 100 µL of DMEM HEPES 10 mM for avoiding evaporation.
13. Check the density of cells using the inverted microscope with the thermostated stage.

Measuring interaction force between T lymphocytes and their target cells 185

14. If the density is very low see notes 1 and 2.
15. Cover the slide with a plastic lid, incubate 15 min at 37 °C.

3.6 Slide installation, immune cell addition (15 min)

1. Clear unbound P815 cells and antibodies in the μ-slide with 4 CV of CD8 culture medium at 37 °C.
2. Fill the Luer with 100 μL of CD8 culture medium at 37 °C (for avoiding bubbles) remove the syringe at the other end.
3. Fill the two Luers of CTL culture medium at 37 °C then put the syringe back.
4. Put the μ-Slide on the thermostated microscope stage.
5. Empty the loading Luer by aspiring slowly the medium using the syringe.
6. Load 50 μL of human CTL suspension in the empty Luer.
7. Fill the Luer with 100 μL of CD8 culture medium.
8. Incubated for 10 min at 37 °C. This incubation time should be optimized for each kind of cell pairs as commented in the note 3.

3.7 Flow rate calibration (5 min)

1. Fill the 60 mL syringe prewarmed at 37 °C with uncolored RPMI medium HEPES 10 mM prewarm at 37 °C and saturated in 5% CO_2 for 1 h.
2. Put the syringe on the syringe pump.
3. Put a 37 °C-thermostating pouch on the top of the syringe.
4. Take an uncoated μ-Slide and set it on the 37 °C-thermostated stage of the inverted microscope.
5. Connect the female Luer of the tubing end to the syringe.
6. Connect the male Luer of the tubing end from the syringe to the μ-Slide Luer.
7. Connect the male Luer of the tubing end connected to the waste plastic bottle to the opposite Luer of the μ-Slide as shown (Fig. 1).
8. Set the parameters of the pump. Syringe diameter, infusion mode, total volume, flow rate 50.
9. Run the pump until the liquid is flowing regularly in the waste bottle then stop.
10. Take a 50 mL conical tube, weight it and replace the waste by this tube.
11. Run the flow rate at 50 for 60 s then stop.
12. Weight the tube.

13. Use the liquid weight to calibrate the data analysis: it was 38.4 mL/min in this example.

3.8 Flow rate gradient and image acquisition (5 min)

1. Fill the 60 mL syringe with uncolored RPMI 10 mM Hepes medium prewarmed at 37 °C and saturated in 5% CO_2 for 1 h.
2. Put the syringe on the syringe pump, orient the eccentric Luer toward the bottom for reducing bubble events.
3. Put a 37 °C-thermostating pouch on the top of the syringe, insulate the tubing.
4. Connect the female Luer of the tubing end to the syringe and purge air by actioning the pump.
5. Fill the loading Luer to the top with 37 °C-prewarmed CTL culture medium using prewarmed tips.
6. Remove delicately and very slowly the syringe with a rotating motion.
7. Fill both loading Luers to the top with 37 °C-prewarmed CTL culture medium using prewarmed tips.
8. Connect delicately the male Luer of the dropping tubing end from the syringe to the overflowed μ-Slide Luer avoiding bubbles.
9. Connect delicately the male Luer of the tubing end connected to the waste plastic bottle or fraction collector to the loading Luer of the μ-Slide.
10. Set the parameters of the program driving the linear flow rate gradient from 0 to 38.4 mL/min in 115 s then hold at 38.4 mL/min. Flow rate depends on the syringe diameter and the pump setting. As described in the note 4, beyond 6 nN, the integrity of cells is challenged. Flow rate gradient should be adapted to the expected force scale.
11. Set the volume per well for the optional fraction collector.
12. Set the image acquisition rate: three images per s (field of view 1100×840 μm).
13. Choose the objective and adjust the focus. See notes 5 and 6.
14. Choose the field of view, preferentially nearby to the flow input along the chamber axis.
15. Note the time since the addition of CD8 cells on P815.
16. Run the program activating the pump flow rate ramp and the image acquisition.
17. Stop the run when all cells are out or when the gradient is over.

The principle of the method is sketched in the Fig. 2.

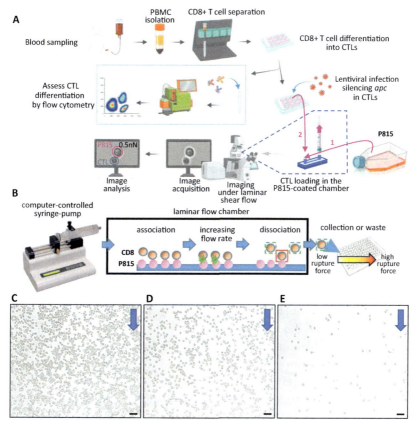

Fig. 2 Graphic summary of the methodology. (A) Workflow of the cells preparation. (B) Principle of the measurement of the interaction force between T lymphocytes and their cell targets using live microscopy and laminar shear flow chamber. Image of the field of view at start (C), 14 s later (D), 30 s later (E). Scale bars (50 μm) are indicated. The blue arrows are indicating the flow direction.

3.9 Image analysis

ImageJ (Schneider et al., 2012), its new release ImageJ2 (Rueden et al., 2017) or the Fiji version (Schindelin et al., 2012) can be used for analyzing image series.

When measuring the adhesion rupture force of CTL to fibronectin, ICAM-1 or VCAM-1 functionalized chamber surface, the automated counting can be used. The diameter of each cell can be extrapolated from the cell surface area or consider the mean value for all of them within an accepted interval of cell surface areas:

1. File > import > image sequence > Folder name >> Filter A-0 >> start 1 >> count - >>Step 10 > scale 100%.
2. Image > type 8 bit.
3. Image > Adjust>threshold clear background (set at the left border of the gaussian distribution of intensity contrast) same threshold for all images.
4. Analyze > set measurement >> area >> stack.
5. Analyze > analyze particles >> size 40–120 pixel^2 >> Circularity 0.8–1 >> show outlines >> display results >> clear results.
6. Save the data file.

For cell-cell contacts manual counting are preferable, the number of images can be reduced. The plugin Cell Counter in early releases has been replaced by the Point tool for a manual counting of cells. A touch-sensitive tablet can be used, more comfortable than mouse and screen.

1. Run ImageJ.
2. File > import > Image sequence, choose the directory, indicate the series head "A-0", choose step 1 for selecting all images, 3 for one roughly per second, or 10 for selecting one image every 10 for reducing the set.
3. Select the Point tool for a manual counting of cells (icon with 5 crosses)
4. Move the image slider to the end.
5. Starting from the end, watching the movie backward image per image, look for the apparition of CTLs. P815s and CTLs were distinguished by their size, shape, and nucleus/cell ratio.
6. Click CTL in contact with P815 when they were not in the previous image and appear in the current one. Ignore CTL alone or P815-CTL pair appearing at the same time
7. Increment the Point tool counter when changing image.
8. At the end click "M" or select Analyze>Measure.
9. Save the data file.

3.10 Data analysis

The flow rate value at the cell–cell rupture event was used to compute the dragging force on the released CTL according to its size (mean diameter of 8 μm), shape (roughly spherical), and density (mean value 1.20 kg/L).

The dragging or Stoke force F_s exerted by the fluid on spherical cells with Stokes radius R_s, is calculated to any point as a linear function of the local flow speed v by the Stokes formula:

$$F_s = 6\,\pi\,\eta\,R_s\,v$$

with the measured medium density 1.0034 kg/L and dynamic viscosity η 0.6998 mPa s at 37 °C. The flow speed v at any point of the chamber (length L, width w and thickness h) can be calculated by applying the Navier-Stokes equation to the flow in the chamber:

$$\rho \left((v_x \, \partial v_x)/\partial x + (v_y \, \partial v_y)/\partial y \right) = -\partial P/\partial x + \rho \, g + \left(\partial^2 v_x/\partial x^2 \quad \partial^2 v_y/\partial y^2 \right)$$

with the pressure P and the gravitational constant g.

At the steady state $\partial v_x / \partial x = 0$ et $\partial^2 v_x / \partial x = 0$ and $\partial v_x / \partial x + \partial v_y / \partial y = 0$

As a result, $\partial v_y / \partial y = 0$ and $v \, y = \text{cste} = 0$.

The Reynolds' number $Re = \rho \, v \, 2 \, R_s$ is below 2000, inertial effects are negligible compared to viscous effects:

$$0 = -\partial P/\partial x + \eta \, \partial^2 v_x/\partial y^2$$
$$\partial^2 v_x/\partial y^2 \quad (1/\eta) \, \partial P/\partial x = (1/\eta) \, P/L$$

By integrating a first time along y: $\partial v_x / \partial y = y \, \Delta P / (\eta \, L) + C1$
and integrating a second time along y: $v_x = y^2 \, \Delta P / 2\eta \, L + y \, C1 + C2$
C1 and C2 are constant values.

The flow speeds at the floor and ceiling contact are: $v_x \, (y = 0) = v_x \, (y = h) = 0$

At the floor $v_x = C2 = 0$ and at the ceiling $v_x = h^2 \, \Delta P / 2 \, \eta \, L + h \, C1 = 0$

Then by replacing C1 in the expression: $v_x = \Delta P / (2 \, \eta \, L) \, (y^2 \quad h \, y)$

The flow rate Q can be calculated using the following expression:

$$Q = \int_{0 \to 1} \int_{0 \to h} v_x \, \partial y \, \partial z = \int_{0 \to h} \Delta P \, w/2\eta \, L \left(y^2 \quad h \, y \right) \partial y$$
$$= -h^3 \, P \, w/12 \, \eta \, L.$$

By replacing $\Delta P = -12 \, \eta \, L \, Q \, / \, (w \, h^3)$, the flow speed is formalized in the following expression:

$$v_x = \frac{6 \, Q}{h^3 \, w} \left(y^2 - h \, y \right)$$

where h and w are the chamber thickness and width respectively and y the distance to the floor. The flow speed calculated throughout the chamber is described in the Fig. 3.

The dragging force exerted by the flow rate on cells should be integrated along the cell height. An approximation has been done for considering the conservation of the spherical shape with the dragging force exerted on cell slices with negligible flow perturbation by the sphere in the vicinity of the glass surface.

The surface area S of a sphere is $4\pi R^2$ and for a b-thick slice of it: $S = 2\pi R b$.

The flow speeds at the middle of 1 μm-slices covering the sphere surface were considered and their contribution to the global dragging force exerted on the sphere was pondered by the ratio of the slice vs sphere areas.

$$F_s = \int_{y=0 \to 2R_s} 6\pi \eta R_s v_{x,y} \left(2\pi R_s b / 4\pi R_s^2\right) dy = \int_{y=0 \to 2R_s} 3\pi \eta v_{x,y} b \, dy$$

F_s is calculated at every step y from the floor (0.05 μm) to the cell top (7.5 μm) (Fig. 3D).

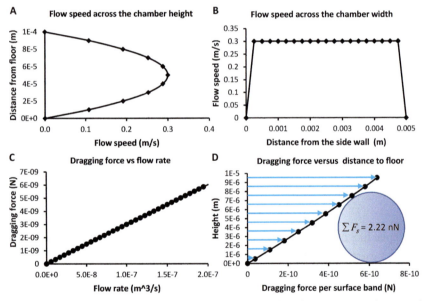

Fig. 3 Data analysis. (A) Flow speed across the chamber height (0.1 mm) at the central axis from top to bottom walls calculated for a flow rate of 100 μm/s. At mid-height the flow speed is 0.29 m/s. (B) Flow speed at mid-height across the chamber width (5 mm) from side to side. (C) Linear relationship between dragging force and flow rate. (D) Averaged dragging force exerted per 1 μm-interval from the chamber floor on cells versus distance to floor. The sum of dragging forces exerted on cell is indicated, corresponding to the mean rupture force for CTL from healthy donor on anti-CD3-coated P815.

Measuring interaction force between T lymphocytes and their target cells

$$F_s = \Sigma_{y=0 \to 2R_s} \, 3 \, \pi \, \eta \, v_{x,y} \, b$$

The cell weight W is determined from the cell density ρ_c and volume V_c and the acceleration of gravity g (9.81 m/s^2) by the formula:

$$W = \rho_c V_c \, g = \frac{-\rho_c * 4}{3} \, \pi \, R_s^3 \, g \, \gamma$$

The buoyancy A applied to the cell immersed in the channel is determined by the formula:

$$A = -\rho_f \, V_c \, g = \frac{\rho_f \, 4}{3} \, \pi \, R_s^3 \, g \, \gamma$$

with the fluid density ρ_f. These two forces are usually represented by a single force:

$$F = \frac{(\rho_f - \rho_c) \, 4}{3} \, \pi \, R^3 \, g \, \gamma$$

Given the size of the cell and the flow speed, this force is negligible.

Analyze data as follows:

1. Open Excel or Prism
2. Import the file saved with ImageJ, with at least columns with the image number and the number of cells in the current image. Plot the cumulative number of cells from start to end.
3. Add a column for the time there is not. The time increment per image is calculated from the total time divided by the number of images.
4. Add a column for the flow rate. The flow rate increment per image is calculated from the calibrated flow rate divided by the time to achieve it.
5. Add as many columns there are slices b in 2 R_s (for example 8 slices of 1 μm in a sphere with a diameter of 8 μm). Calculate F_s at mid height (y) of the slice (b) with the following formula for the chamber of width w and thickness h and the current flow rate Q (Fig. 3D):

$$F_s = \frac{18 \, \pi \, \eta \, b \, Q}{h^3 \, w} \, (y^2 - h \, y)$$

6. Sum the F_s per slice columns in a total dragging force column.
7. Plot the cumulative number of cells vs the dragging force values (Fig. 4B).
8. Represent the quartile values of the dragging forces among the cell population using a whisker-box plot (Fig. 4C).

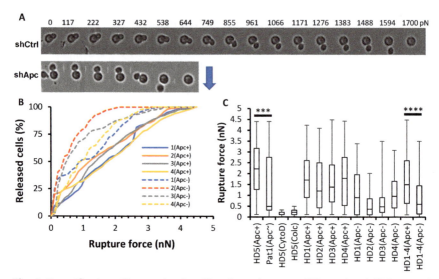

Fig. 4 Quantification of interaction breaking forces between CTLs and anti-CD3–coated P815 cells. Control or *apc*-silenced human CTLs were set to interact with anti-CD3–coated P815 tumor target cells previously adhered to a fibronectin-coated laminar flow chamber for 10 min at 37 °C. Then, a laminar flow of PBS increasing from 0 to 38.4 mL/min was applied through the chamber for 115 s using a computer-driven syringe pump and synchronized with image acquisition (three images per s). The blue arrow is indicating the flow direction. (A) Representative images showing the sequential steps of CTL detachment from the target cell. CTLs are distinguished from target cells by their smaller size, cytoplasm, and light refringence. Top line, shCtrl (original magnification 3250), and bottom line, shApc-treated cells (original magnification 3300). Force in nN corresponding to each image is depicted. (B) Four healthy donors (HD1–4) treated with shCtrl (Apc$^+$, $n=684, 671, 1189, 1201$ CTLs) or with shApc (Apc$^-$, $n=576, 239, 124, 152$ CTLs) were analyzed. Cumulative plots of cell–cell rupture events versus dragging force combining data from the four donors shCtrl and corresponding shAPC treated CTLs. (C) Whiskers represent minimum and maximum values, whereas boxes represent second and third quartiles framing the median. Quantification of rupture forces of interactions between CTL and anti-CD3–coated P815 cells from left to right: untreated CTL from healthy donor (HD5) and apc-mutated patient (Pat1), cytochalasin-D (HD5 CytoD) and colchicine-treated (HD5 Colc) CTL, shCtrl-treated CTL (Apc$^+$) and shApc-treated CTL (Apc) from HD1–4, and the combined data Apc$^+$ and Apc$^-$ from the four donors with a significance determined by Mann–Whitney *U* test: *P*, 0.001 (***) and 0.0001 (****) indicated at the top of compared datasets.

4. Results

The polarity regulator and tumor suppressor *adenomatous polyposis coli* (APC) protein is involved in the cytoskeleton organization of the synapse (Aguera-Gonzalez et al., 2017; Cuche et al., 2023; Juzans et al., 2020) monitoring the attachment of cytotoxic T-cell (CTL) to targeted cells, and their migration (Mastrogiovanni et al., 2022). The *apc* gene mutations are associated with familial adenomatous polyposis and the initiation of human colorectal tumors. APC mutations or truncates alter intestinal epithelium differentiation and induce tumor progression (Béroud & Soussi, 1996; McCartney & Näthke, 2008; Moser, Pitot, & Dove, 1990). To obtain further insight, we studied the effect of *apc* gene silencing of CTL on the stability of their interaction with tumor target cells, we used cells from P815 mouse mastocytoma cell line coated with anti-CD3 antibodies as model of tumor target cell and human in vitro differentiated CTLs (Juzans et al., 2020).

By using the laminar flow chamber described above to generate dragging force on cell pairs, we measured the force required to rupture the cell-cell interactions between CTL and anti-CD3-coated P815 cells. CTLs interacting with immobilized P815 cells were stretched by this increasing dragging force until rupture of the contact and release of the CTL into the stream (Fig. 4A). CTL–P815 cell rupture forces ranged from 0 to 4.5 nN with half of values between 0.5 and 2.5 nN suggesting a wide variety of synapse occurrence kinetics and/or complexity (Fig. 4C). Average rupture force was 2.22 nN for untreated CTL control (HD), 0.11 nN for Cytochalasin-D-treated CTL (inhibition of actin polymerization), 0.22 nM for Colchicine-treated CTL (inhibition of microtubule assembly), 0.49 nN for patient's CTL (identified truncated *apc* gene, associated with familial adenomatous polyposis) and 1.52 nN for gene-silencing control and 0.67 nN *apc*-silenced cells, indicating that this interacting force is decreased by loss impaired APC expression and/or function (Fig. 4C).

Since the number of cells loaded in the chamber and visible in the field of view of the microscope may vary from one experiment to another, the following precaution was used. In order to give the same weight to each of the four experiments performed on different cell numbers in the combined data whisker-boxes shCtrl ($n = 3745$) and shApc ($n = 1091$), cells released per force interval were scaled to a total number of 250 cells per experiment, 1000 cells after adding the four sets.

Interestingly, cumulative plots of cell–cell rupture events versus dragging force show linear distributions for control cells and hyperbolic distributions for *apc*-silenced cells (Fig. 4B), reflecting the homogeneous activation state of the cell population: hyperbolic more homogeneous, linear more diverse. Of note is that although equal cellular input of shCtrl and shApc cells was applied to the chamber, the number of CTLs forming initial conjugates with P815 cells before applying flow pressure was lower in shApc-treated cells in most experiments. This further reflects the lesser capacity of *apc*-silenced cells to initially stabilize conjugates with P815 target cells.

5. Concluding remarks

If *apc* gene mutations or truncates have been associated with familial adenomatous polyposis, these patients may also feature other mutated genes. Our measurements shows that *apc* gene silencing are sufficient for impeding the adhesion force weakening the synapse. Similar effects were observed when actin and microtubule cytoskeleton were perturbed by drug inhibitors. Further experiments have shown that kinetics of adhesion force increase is also affected by *apc* gene alteration, suggesting that CTL contact has to be longer with the tumor cells to stop the CTL motion. Such effect should thus favor tumor cells escape to immune surveillance.

Quantifying the rupture force of cell-cell interactions using laminar shear flow enables efficient differentiation of subtypes or activation states between cell populations or distribution within them. It enables to screen the effects of medium-added inhibitors or activators on cell-cell interactions or determine the involvement of expressed proteins using gene silencing as illustrated in this study. Kinetic measurements of synapse formation and reinforcement are measured as a function of contact time, controlled by incubation or by rolling immune cells at very low flow rates onto targets. A large number of factors can affect the kinetics and strength of cell-cell interaction, and even more so the quantification of their disruption, as indicated in the following notes. Measured values are thus indicative of the order of magnitude. The comparison of relative values between one experiment and another is the main strength of this method when carried out under conditions that are, if not identical, at least very close.

This robust method thus offers significant tools in studying intercellular forces, synapse dynamics, and their implications in diverse processes, which are crucial for understanding cellular functions in immunology research.

6. Perspectives

Image series analysis of the disruption of cell interaction with a surface functionalized with fibronectin, ICAM-1 or VCAM-1 are easy and automatable with ImageJ (Mastrogiovanni et al., 2022). Analysis of the disruption of interaction between two cells is much more complicated, and is requiring visual and manual validation. A deep learning-based program for analyzing pairs of cells in image series using artificial intelligence is currently under development.

Additional research studies focusing on the interaction strength of immune cells (CD4 T lymphocytes, natural killers and macrophages) with different solid tumor cell targets have been carried out and will be published.

Notes

1. The target cells were placed in the flow chamber under culture conditions. If the target cells are not adherent, it is possible to use poly-lysine to reinforce the adhesion. Poly-lysine excess will then have to be neutralized with BSA before adding the T lymphocytes, 15 min after the first cells.

2. If the target cells are adherent, they should be placed in culture conditions for several hours or even overnight to promote their spreading and attachment by periodically renewing the warmed medium (37 °C) pre-equilibrated with dissolved gas (5% CO_2) in a bottle with a gas-permeable cap. If the cells do not adhere to glass, it is possible to stick the μ–Slide in a polystyrene Petri dish or a polystyrene slide treated for cell culture. The higher the cell density, the higher the number of pairs to observe. However, if they are too tightly packed there may be a risk that the T cells interact with more than one target cell. Target cells could be surface-adhered and adopt flat to spherical shapes.

   As soon as the target cells are firmly adhered, T cells cultured for at least 30 min at 37 °C in CO_2 under the test conditions of activation (cytokines) or preparation (medium additives, binding or inhibiting antibodies) can be injected into the flow chamber.

3. Variation of incubation time of pair occurrences has to take into account the kinetics of the interaction force increase and thus the number of adhesion molecules at the interface. This is an important factor when the T cells are injected while cell targets are immobilized on the chamber floor. The time it takes for the T cell to make a contact with the floor-immobilized target is not exactly the same for all cells whatever the T cells are carried by a very slow flow rate rolling them at the surface of glass or loaded through the chamber at high rate then sediment to the floor. This time depends on interactions, often weak, provided by specific surface receptors out of integrins. This time of contact prior the exercise of force for pulling the target and the T cell apart, should be taken into account as adhesion force will increase with time because of an increasing number of integrin pairs contributing to this adhesion.

4. We have noticed that beyond a force of 6 nN, a fraction of the collected cells does not grow well, suggesting the rupture of their membrane and the loss of their cell integrity.

5. The use of a fluorescence microscope and appropriate specific labelling will allow to differentiate cells, identify their polarity, and observe the organization of surface proteins or the structuring of the cytoskeleton or synapse in real-time.

6. A low microscope magnification allows a large field of view. A higher magnification is possible if the working distance of objective is compatible with the thickness of the bottom glass slide. We used 1 mm-thick glass slide to keep a focus on cells and avoid the chamber slide floor deformation due to the flow pressure. In comparison we lost the focus on cells quickly when using a 0.15 mm-thick glass slide.

Globally these experiments require some organization and training. If quantification of rupture forces of interactions between spherical cells on flat functionalized surface are straightforward, it is much more delicate when considering cell pairs, and the data analysis can be tedious. This protocol is proposed to compare rupture forces as a function of conditions, and therefore to discuss relative values rather than raw values. Moreover, to obtain accurate data, there are a number of crucial points to take into account.

1. Air bubbles in the chamber must be avoided at all stages, particularly when the μ-Slide is placed on the microscope. Any bubbles will drag all the cells with them into the flow. Indeed, this is the way we collect the last flow resistant cells.

2. Cells should never be allowed to fall below 32 °C in the 30 min prior to force measurements, to avoid affecting adhesion kinetics and strength.

3. The rupture force depends on the time lag between cell attachment and detachment, as the formation of immunological synapses is a progressive mechanism. However, this mechanism is slow, and by reproducing the measurement protocol with the same incubation times, it is possible to compare the results from one experiment to another

Several additional facts may affect data analysis.

1. Cell adhesion to the floor can flatten them, altering their height and diverging their shape from a sphere, as in the case of P815. This is not the case if the adhesion time is short (<15 min) and if the adhesion experiments are conducted with another cell, for instance CTL binding to P815.

2. CTL adhesion with fibronectin should be reduced by limiting the time of contact prior to measurement. CTL mobility in the flow should be checked from oscillation and rotation around the P815 attachment to confirm the absence of interaction with fibronectin. If there is binding only to P815, the CTL membrane deforms to form a long, cell-sized

filipode. At the moment of rupture, the far away P815 no longer really protects the CTL, which is subjected to laminar flow.

3. The higher the flow, the greater the pressure exerted on the cells, causing them to flatten out according to the rigidity conferred by their cytoskeleton. The resulting loss of height will reduce the friction force. On the other hand, when a CTL binds to a P815, part of the CTL is protected from liquid friction. However, deformation of the P815 caused by its elongation under the increasing flow speed reduces this effect.

4. The flow near the glass surface is perturbed by the cell (Goldman, Cox, & Brenner, 1967a, 1967b) that might affect the calculation of the friction force near the glass surface.

5. The dragging force has been considered in this method as a sum of forces on cell slices. Other methods have been published earlier (Olivier & Truskey, 1993; Pierres, Tissot, Malissen, & Bongrand, 1994; Chapman & Cokelet, 1996). This method of data analysis is not better for raw rupture force measurements but robust and convenient for real time processing as well post-acquisition of relative rupture force from one cell pair to another and rupture force distribution from one experiment to another.

Ethics statement

The studies involving human participants were reviewed and approved by Comité de Protection des Personnes, Île de France-1. Protocol N° 2010-déc. 12,483 for healthy subjects and N° 2018-mai-14,852 for FAP subjects. The patients/participants provided their written informed consent to participate in this study.

Acknowledgments

We thank the Association Polyposes Familiales France and the Institut Pasteur ICAReB Biological Resources core facility for their key contribution to the recruitment of FAP patients and healthy volunteers and the corresponding blood samples. We also thank all the volunteers for their participation in this project. We thank the ICAReB Biological Resources core facility team, Institut Pasteur from healthy donors and patients. This work was supported by grants from La Ligue Nationale contre le Cancer and Equipe Labellisée Ligue 2018 and institutional grants from Institut Pasteur and INSERM. The Diagnostic Test Innovation & Development is supported by the grant SESAME from the French public investment bank BPI), region Ile-de-France and the Institut Pasteur and the grant from the Agence Nationale de la Recherche (ANR 2021 CE33 0003 03 OPTOBOT). NI and MJ have been funded by the ANR grant (OPTOBOT) and an Allocation de Recherche Doctorale from La Ligue Nationale Contre le Cancer respectively.

Declaration of interest

The authors declare that they have no conflict of interest.

References

Aguera-Gonzalez, S., Burton, O. T., Vazquez-Chavez, E., Cuche, C., Herit, F., Bouchet, J., et al. (2017). Adenomatous polyposis coli defines Treg differentiation and anti-inflammatory function through microtubule-mediated NFAT localization. *Cell Reports, 21*, 181–194. https://doi.org/10.1016/j.celrep.2017.09.020.

Ashkin, A., Dziedzic, J., & Yamane, T. (1987). Optical trapping and manipulation of single cells using infrared laser beams. *Nature, 330*, 769–771. https://doi.org/10.1038/330769a0.

Benoit, M., Gabriel, D., Gerisch, G., & Gaub, H. E. (2000). Discrete interactions in cell adhesion measured by single-molecule force spectroscopy. *Nature Cell Biology, 2*, 313–317. https://doi.org/10.1038/35014000.

Benoliel, C., Capo, J. L., & Mege and P. Bongrand. (1994). Measurement of the strength of cell-cell and cell-substratum adhesion with simple methods. *Studying Cell Adhesion*, 81–92. Springer-Verlag.

Béroud, C., & Soussi, T. (1996). APC gene: Database of germline and somatic mutations in human tumors and cell lines. *Nucleic Acids Research, 24*, 121–124. https://doi.org/10.1093/nar/24.1.121.

Boettiger, D. (2007). Quantitative measurements of integrin-mediated adhesion to extracellular matrix. *Methods in Enzymology, 426*, 1–25. https://doi.org/10.1016/S0076-6879 (07)26001-X.

Chapman, & Cokelet. (1996). Model studies of leukocyte-endothelium-blood interactions: I. The fluid flow drag force on the adherent leukocyte. *Biorheology, 33*(2), 119–138. https://doi.org/10.1016/0006-355X(96)00011-X.

Cuche, C., Mastrogiovanni, M., Juzans, M., Laude, H., Ungeheuer, M. N., Krentzel, D., et al. (2023). T cell migration and effector function differences in familial adenomatous polyposis patients with APC gene mutations. *Frontiers in Immunology, 18*(14), 1163466. https://doi.org/10.3389/fimmu.2023.1163466.

Dustin, M. L., & Springer, T. A. (1989). T-cell receptor cross-linking transiently stimulates adhesiveness through LFA-1. *Nature, 341*, 619–624. https://doi.org/10.1038/341619a0.

Edelstein, A. D., Tsuchida, M. A., Amodaj, N., Pinkard, H., Vale, R. D., & Stuurman, N. (2014). Advanced methods of microscope control using μManager software. *Journal of Biological Methods, 1*(2), e11. https://doi.org/10.14440/jbm.2014.36.

Favre-Bulle, I. A., & Scott, E. K. (2022). Optical tweezers across scales in cell biology. *Trends in Cell Biology, 32*(11), 932–946. https://doi.org/10.1016/j.tcb.2022.05.001.

Friedland, J. C., Lee, M. H., & Boettiger, D. (2009). Mechanically activated integrin switch controls alpha5-beta1 function. *Science, 323*, 642–644. https://doi.org/10.1126/science.1168441.

Garcia, A. J., Huber, F., & Boettiger, D. (1998). Force required to break alpha5beta1 integrin-fibronectin bonds in intact adherent cells is sensitive to integrin activation state. *The Journal of Biological Chemistry, 273*, 10988–10993. https://doi.org/10.1016/S0142-9612(97)00042-2.

Goldman, A. J., Cox, R. G., & Brenner, H. (1967a). Slow viscous motion of a sphere parallel to a plane wall- I motion through a quiescent fluid. *Chemical Engineering Science, 22*(4), 637–651. https://doi.org/10.1016/0009-2509(67)80047-2.

Goldman, A. J., Cox, R. G., & Brenner, H. (1967b). Slow viscous motion of a sphere parallel to a plane wall-II Couette flow. *Chemical Engineering Science, 22*(4), 653–660. https://doi.org/10.1016/0009-2509(67)80048-4.

Gosse, C., & Croquette, V. (2002). Magnetic tweezers: Micromanipulation and force measurement at the molecular level. *Biophysical Journal, 82*, 3314–3329. https://doi.org/10.1016/S0006-3495(02)75672-5.

Hogan, B., Babataheri, A., Hwang, Y., Barakat, A. I., & Husson, J. (2015). Characterizing cell adhesion by using micropipette aspiration. *Biophysical Journal, 109*, 209–219. https://doi.org/10.1016/j.bpj.2015.06.015.

Juzans, M., Cuche, C., Di Bartolo, V., & Alcover, A. (2023). Imaging polarized granule release at the cytotoxic T cell immunological synapse using TIRF microscopy: Control by polarity regulators. *Methods in Cell Biology, 173*, 1–13. https://doi.org/10.1016/bs.mcb.2022.07.016.

Juzans, M., Cuche, C., Rose, T., Mastrogiovanni, M., Bochet, P., Di Bartolo, V., et al. (2020). Adenomatous polyposis coli modulates actin and microtubule cytoskeleton at the immunological synapse to tune CTL functions. *Immunohorizons, 4*, 363–381. https://doi.org/10.4049/immunohorizons.2000044.

Kamsma, D., Bochet, P., Oswald, F., Ablas, N., Goyard, S., Wuite, G. J. L., et al. (2018). Single-cell acoustic force spectroscopy (scAFS): Resolving kinetics and strength of T-cell adhesion to fibronectin. *Cell Reports, 2018*(24), 3008–3016. https://doi.org/10.1016/j.celrep.2018.08.034.

Kamsma, D., Creyghton, R., Sitters, G., Wuite, G. J., & Peterman, E. J. (2016). Tuning the music: Acoustic force spectroscopy (AFS) 2.0. *Methods, 105*, 26–33. https://doi.org/10.1016/j.ymeth.2016.05.002.

Kaplanski, G., Farnarier, C., Tissot, O., Pierres, A., Benoliel, A. M., Alessi, M. C., et al. (1993). Granulocyte-endothelium initial adhesion: Analysis of transient binding events mediated by E-selectin in a laminar shear flow. *Biophysical Journal, 64*, 1922–1933. https://doi.org/10.1016/S0006-3495(93)81563-7.

Kim, S. H., Chegal, W., Doh, J., Cho, H. M., & Moon, D. W. (2011). Study of cell-matrix adhesion dynamics using surface plasmon resonance imaging ellipsometry. *Biophysical Journal, 100*, 1819–1828. https://doi.org/10.1016/j.bpj.2011.01.033.

Kinashi, T. (2012). Overview of integrin signaling in the immune system. *Methods in Molecular Biology, 757*, 261–278. https://doi.org/10.1007/978-1-61779-166-6_17.

Lehenkari, P. P., & Horton, M. A. (1999). Single integrin molecule adhesion forces in intact cells measured by atomic force microscopy. *Biochemical and Biophysical Research Communications, 259*, 645–650. https://doi.org/10.1006/bbrc.1999.0827.

Li, F., Redick, S. D., Erickson, H. P., & Moy, V. T. (2003). Force measurements of the alpha5beta1 integrin-fibronectin interaction. *Biophysical Journal, 84*, 1252–1262. https://doi.org/10.1016/S0006-3495(03)74940-6.

Lu, H., Koo, L. Y., Wang, W. M., Lauffenburger, D. A., Griffith, L. G., & Jensen, K. F. (2004). Microfluidic shear devices for quantitative analysis of cell adhesion. *Analytical Chemistry, 76*, 5257–5264. https://doi.org/10.1021/ac049837t.

Mastrogiovanni, M., Vargas, P., Rose, T., Cuche, C., Esposito, E., Juzans, M., et al. (2022). The tumor suppressor adenomatous polyposis coli regulates T lymphocyte migration. *Science Advances, 8*(15). https://doi.org/10.1126/sciadv.abl5942. eabl5942.

McCartney, B. M., & Näthke, I. S. (2008). Cell regulation by the protein Apc as master regulator of epithelia. *Current Opinion in Cell Biology, 20*, 186–193. https://doi.org/10.1016/j.ceb.2008.02.001.

Merkel, R., Nassoy, P., Leung, A., et al. (1999). Energy landscapes of receptor–ligand bonds explored with dynamic force spectroscopy. *Nature, 397*, 50–53. https://doi.org/10.1038/16219.

Moser, A. R., Pitot, H. C., & Dove, W. F. (1990). A dominant mutation that predisposes to multiple intestinal neoplasia in the mouse. *Science, 247*, 322–324. https://doi.org/10.1126/science.2296722.

Olivier, & Truskey. (1993). A numerical analysis of forces exerted by laminar flow on spreading cells in a parallel plate flow chamber assay. *Biotechnology and Bioengineering, 42*(8), 963–973. https://doi.org/10.1002/bit.260420807.

Pierres, A., Tissot, O., & Bongrand, P. (1994). Analysis of the motion of cells driven along an adhesive surface by a laminar shear flow. *Studying Cell Adhesion*, 157–173. https://doi.org/10.1007/978-3-662-03008-0_11. Springer-Verlag.

Pierres, A., Tissot, O., Malissen, B., & Bongrand, P. (1994). Dynamic adhesion of CD8-positive cells to antibody-coated surfaces: The initial step is independent of microfilaments and intracellular domains of cell-binding molecules. *The Journal of Cell Biology*, *125*, 945–953. https://doi.org/10.1083/jcb.125.4.945.

Reyes, C. D., & Garcia, A. J. (2003). A centrifugation cell adhesion assay for high-throughput screening of biomaterial surfaces. *Journal of Biomedical Materials Research. Part A*, *67*, 328–333. https://doi.org/10.1002/jbm.a.10122.

Robert, P., Limozin, L., van der Merwe, P. A., & Bongrand, P. (2021). CD8 Co-Receptor Enhances T-Cell Activation without Any Effect on Initial Attachment. *Cell*, *10*, 429. https://doi.org/10.3390/cells10020429.

Rueden, C. T., Schindelin, J., Hiner, M. C., DeZonia, B. E., Walter, A. E., Arena, E. T., et al. (2017). ImageJ2: ImageJ for the next generation of scientific image data. *BMC Bioinformatics*, *18*(1). https://doi.org/10.1186/s12859-017-1934-z.

Schaffer, E., Norrelykke, S. F., & Howard, J. (2007). Surface forces and drag coefficients of microspheres near a plane surface measured with optical tweezers. *Langmuir*, *23*, 3654–3665. https://doi.org/10.1021/la0622368.

Schindelin, J., Arganda-Carreras, I., Frise, E., Kaynig, V., Longair, M., Pietzsch, T., et al. (2012). Fiji: An open-source platform for biological-image analysis. *Nature Methods*, *9*(7), 676–682. https://doi.org/10.1038/nmeth.2019.

Schneider, C. A., Rasband, W. S., & Eliceiri, K. W. (2012). NIH image to ImageJ: 25 years of image analysis. *Nature Methods*, *9*(7), 671–675. https://doi.org/10.1038/nmeth.2089.

Shao, J. Y., & Hochmuth, R. M. (1996). Micropipette suction for measuring picoNewton forces of adhesion and tether formation from neutrophil membranes. *Biophysical Journal*, *71*, 2892–2901. https://doi.org/10.1016/S0006-3495(96)79486-9.

Sitters, G., Kamsma, D., Thalhammer, G., Ritsch-Marte, M., Peterman, E. J., & Wuite, G. J. (2015). Acoustic force spectroscopy. *Nature Methods*, *12*, 47–50. https://doi.org/10.1038/nmeth.3183.

Wang, X., Chen, S., Kong, M., Wang, Z., Costa, K. D., & LI, R.A., Sun, D. (2011). Enhanced cell sorting and manipulation with combined optical tweezer and microfluidic chip technologies. *Lab on a Chip.*, 3656–3662. https://doi.org/10.1039/C1LC20653B.

CHAPTER NINE

Isolation and characterization of primary NK cells and the enrichment of the KIR2DL1[+] population

Batel Sabag[†], Abhishek Puthenveetil[†], and Mira Barda-Saad[*]
The Mina and Everard Goodman Faculty of Life Sciences, Bar-Ilan University, Ramat-Gan, Israel
[*]Corresponding author: e-mail address: mira.barda-saad@biu.ac.il

Contents

1. Introduction 202
2. Materials 203
   2.1 Common disposables 203
   2.2 Common reagents 204
   2.3 Common equipment 205
   2.4 Cell lines 205
3. Methods 205
   3.1 Reagent preparation 205
   3.2 Isolation for KIR2DL1 positive population 209
4. Concluding remarks 210
Acknowledgments 211
References 211

Abstract

Natural killer (NK) cells are cytotoxic innate lymphoid cells that play critical roles in the mitigation of viral infections and cancer through the secretion of cytolytic granules and immunomodulatory cytokines. Abnormalities in NK function can lead to viral infections, autoimmunity, and cancer. The current protocol provides an NK isolation technique to study the signaling pathways downstream to the Killer cell immunoglobulin-like receptors (KIR) that serve as key human NK cell function regulators. This procedure enables investigating mechanisms specific to individual KIRs to improve our understanding of NK cell function in health and disease.

[†] These authors contributed equally to this work.

1. Introduction

Natural killer (NK) cells are a subset of lymphocytes that play a crucial role in the immune system's defense against tumorigenic cells, pathogens, and virally infected cells. They serve as a major line of defense by directly identifying and eliminating target cells without the need for prior sensitization. Unlike T-lymphocytes, which rely on specific antigens presented by the Major Histocompatibility Complex (MHC), NK cells utilize surface receptors to directly recognize their targets.

The ability of NK cells to distinguish between healthy and transformed cells is based on the "missing self" principle (Anfossi et al., 2006; Raulet, 2006). NK cells recognize and eliminate target cells that fail to express self-MHC class I molecules. They express a diverse array of germline-encoded activating and co-activating receptors. Additionally, they possess inhibitory receptors known as Killer-cell Immunoglobulin-like Receptors (KIRs). These inhibitory receptors, such as KIR2DL1 in humans, interact with MHC-I molecules expressed on various cell types, thereby regulating NK cell effector functions (Anfossi et al., 2006; Yawata et al., 2008).

In humans, KIR molecules belong to the immunoglobulin superfamily, while in rodents and other species, the Ly49 protein family fulfills a similar role. These inhibitory receptors distinguish between different MHC-I allelic variants, enabling the detection of virally infected or transformed cells.

A complex interplay of activating and inhibitory signals tightly regulates the activity of NK cells. The balance between these signals determines whether NK cell activation or inhibition occurs (Long, Sik Kim, Liu, Peterson, & Rajagopalan, 2013; Vivier, 2008). Activation involves the phosphorylation of key signaling proteins, while inhibition is mediated by the phosphatase SHP-1, which serves as a negative regulator of upstream activation.

Understanding the mechanisms that ensure NK cell tolerance and the dynamics of SHP-1 activity is crucial. To investigate these aspects, researchers often employ NK cell lines such as YTS, which express specific inhibitory receptors like KIR2DL1, and target cells expressing different MHC-I alleles. Primary NK (pNK) cells, isolated from peripheral blood, are used for more physiologically relevant studies. These pNK cells express inhibitory receptors like KIR2DL1 and can be tested against target cells expressing or lacking specific MHC-I molecules. The target cells are 721.221-HLA-Cw4 cells that express the HLA of allotype Cw4, the ligand for the KIR2DL1 inhibitory receptor, and thus induce an inhibitory interaction. In contrast, the 721.221-HLA-Cw7 cells express HLA of allotype Cw7 that is not recognized by the KIR2DL1 receptor, thus results in NK cell activation (Ben-Shmuel et al., 2022; Ben-Shmuel, Biber, Sabag, & Barda-Saad, 2021; Biber et al., 2021; Matalon et al., 2016; Matalon et al., 2018; Sabag et al., 2022).

KIRs are a family of type I transmembrane glycoproteins expressed on the surface of NK cells and a minor population of T cells. They exhibit a diverse repertoire, with multiple alleles described for certain KIRs. The role of inhibitory KIRs in educating NK cells has been extensively studied (Lanier, 2008). The expression of KIRs on NK cells initially occurs in a stochastic manner. However, NK cells undergo an education process mediated by inhibitory receptors. This process allows NK cells to mature and modulate KIR expression, striking a balance between self-tolerance and effector functions (Yu et al., 2007).

The density of KIR receptors does not influence NK cell education at the single-cell level (Béziat et al., 2013). However, individuals with multiple KIR gene copies exhibit higher frequencies of responding cells, suggesting heightened overall responsiveness. Resting NK cells utilize preformed lytic granules to exert their cytolytic effects on target cells, indicating the need for precise control mechanisms. Compared to T-cells, NK cells possess advantageous cytolytic capabilities, making them a promising candidate for cancer immunotherapy (Hsu et al., 2016). Extensive research is being conducted to uncover the functionality of NK cells and their inhibitory receptors, such as KIRs, to harness their therapeutic potential fully.

To facilitate the investigation of NK cell functions, a feasible method has been developed to isolate and enrich NK cells from peripheral blood mononuclear cells. This method allows for the specific isolation of particular KIR populations, enabling researchers to delve deeper into the role of specific KIRs and better understand their underlying functions in NK cell biology.

2. Materials
2.1 Common disposables

15 mL screw cap tubes
50 mL screw cap tubes
5 mL serological pipettes
10 mL serological pipettes
25 mL serological pipettes

Sterile pipette tips

250 mL Flask

Falcon® Round-Bottom Polystyrene Tubes, 5 mL

Syringe

Pasteur pipette

0.22 μm filter

30 μm nylon mesh

96 well plate

0.22 μm vacuum filter bottle

2.2 Common reagents

Hank's Balanced Salt Solution, HBSS, no calcium, no magnesium, no phenol red (02-018-1A-500 ML, Biological Industries).

Density gradient medium for the isolation of mononuclear cells, Lymphoprep™ (#1861-1, Serumwerk Benburg).

Dulbecco's Phosphate Buffered Saline, DPBS 10× (#D1408-500ML, Sigma-Aldrich).

RPMI-1640 Medium (#R8758-500ML, Sigma-Aldrich).

Dulbecco's modified Eagle's medium (DMEM) (#01–052-1A, Biological Industries).

Ham's F-12 Nutrient mix (#21765029-500ML, Gibco).

Human serum (#H4522-100ML, Sigma-Aldrich).

L-glutamine (#G7513-100ML, Sigma-Aldrich).

Penicillin & streptomycin solution (#P4333-100ML, Sigma-Aldrich).

Non-essential amino acids, MEM EAGLE 100× (#01–340-1B, Biological Industries).

Sodium pyruvate (#S8636-100ML, Sigma-Aldrich).

Fetal Bovine serum, FBS (#F2442-500ML, Sigma-Aldrich).

Phytohemagglutinin (PHA) (#L8902-5MG, Sigma-Aldrich).

Recombinant Human IL-2 (CYT-209).

Recombinant Human IL-15(CYT-230).

Trypan blue (#93595-250ML, Sigma-Aldrich).

Ethylenediaminetetraacetic acid tripotassium salt dihydrate, EDTA.

EasySep™ Human NK Cell Enrichment Kit (#19055, StemCell Technologies Inc).

EasySep™ Human PE Positive Selection Kit II (#17664, StemCell Technologies Inc).

Antibodies

CD158a/h (KIR2DL1/DS1) Antibody, anti-human, PE (#130-099-209, Miltenyi Biotec).
FITC anti human CD56 (NCAM), clone HCD56 (#318303, BioLegend).
PE-Cy5 Mouse anti human CD3 (#555334 BD Biosciences).

2.3 Common equipment

Rubber pump.
Scissor.
Laboratory centrifuge.
Incubator set at 37 °C with 5% CO2 in air and ≥95% humidity.
Vortex.
Pipette-aid.
Pipettors.
EasySep™ Magnet (Catalog #18000).
Standard light microscope for cell counting.
Flow cytometry.

2.4 Cell lines

Two variants of the human 721.221 B-lymphoblastoid cell line, 721.221 target cells that overexpress the HLA–Cw4, a cognate MHC-I ligand for the KIR2DL1 (721.221-HLA-Cw4), and 721.221 target cells that overexpress irrelevant HLA-Cw7 (721.221-HLA-Cw7) molecule. These cells were kindly provided by Prof. Ofer Mandelboim (Department of Microbiology and Immunology, Faculty of Medicine, Hebrew University of Jerusalem, Israel). 721.221-HLA-Cw4/7 cells were cultured in RPMI supplemented with 10% FBS, 2 mM L-glutamine, 1% sodium pyruvate, 50 μg/mL penicillin and streptomycin.

3. Methods

3.1 Reagent preparation

EDTA 0.5 M solution

Prepare stock solution (3.2 μg/mL) by dissolving 9.306 g of EDTA powder in 30 mL of DDW adjust the pH to pH = 8 with NaOH and add DDW to final volume of 50 mL.

Separation buffer

- o   49 mL DPBS 1×
- o   1 mL FCS (2% in the final volume)
- o   100 µL EDTA 0.5 M (1 mM in the final volume)

Filter the buffer and keep inside the Hood (for degassing).

Prepared fresh every time (but can be stored at 4 °C – not more than 2 weeks).

Primary NK cell media preparation

- o   60% Dulbecco's modified Eagle's medium (DMEM)
- o   25% F-12 medium supplemented
- o   10% human serum
- o   2 mM L-glutamine
- o   50 µg/mL penicillin, 50 µg/mL streptomycin
- o   1% non-essential amino acids
- o   1% sodium pyruvate.

Filter the medium using 0.22 µm vacuum filter bottle.

Expand the cells with medium containing 1 µg/mL of PHA, 400 U/mL rhu IL-2 and 400 U/mL rhu IL-15.

Isolation of mononuclear cells (MNCs) from human peripheral blood by density gradient centrifugation

1. Prepare ten 15 mL tubes and add to each one of the tubes 3 mL Lymphoprep™ (work in dark)
2. Wash the blood bag and scissors with 70% ethanol, and cut the blood bag
3. Pour the blood to a flask and add HBSS 1:1 (50 mL blood +50 mL HBSS)
4. Mix the blood and the HBSS to homogeneous solution with 10 mL pipette
5. Add slowly 10 mL diluted blood on 3 mL Lymphoprep™
6. Centrifuge 30 min 1700 rpm without break for the Lymphoprep™ gradient
7. Vacuum careful the plasma until 7 mL mark on the tube, it will help to reach easily to the PBMCs
8. Collect the mononuclear cells with ring with pasteur pipette into a 50 mL tube with 20 mL RPMI medium.
9. Add RPMI medium to total volume of 50 mL and centrifuge for 10 min 1600 rpm with break
10. Take out the supernatant and add 20 mL RPMI medium and pipet with a 10 mL pipette
11. Add RPMI medium to the final volume of 50 mL and Centrifuge for 10 min 1200 rpm

12. Take out the supernatant and add 20 mL Complete medium and pipet well with a 10 mL pipette
13. Add complete medium to the final volume of 50 mL, transfer trough 30 μm nylon mesh into new 50 mL tube
14. Take 10 μL for counting and mix with 90 μL Trypan blue, count only big round white cells*
    * ≅500–1000*10^6 PMBCs were obtained from 50 mL blood.

Isolation of primary Natural killer cells by negative selection

1. Transfer PBMCS into a tube.
2. Centrifuge for **7 min** 1400 rpm (400 g).
3. Vacuum supernatant (don't disturb the pellet).
4. Add 1.5 mL separation buffer on the pellet and pipette carefully. Measure the volume of the cell solution and transfer into FACS tube (Falcon® Round-Bottom Polystyrene Tubes, 5 mL). Add the desired volume to reach to the final volume of 2 mL. Work delicate, try to avoid bubbles caused by pipette.
5. Use kit stem cell technologies NK enrichment-**negative selection**.
6. Add easy Human NK cell enrichment 100 μL of cocktail (50 μL/mL cells). Mix well with 1 mL pipette.
7. Incubate at R.T for **10** min.
8. Vortex the magnetic beads for **30s**.
9. Add 200 μL magnetic beads to cells (100 μL/mL). Mix well with 1 mL pipette.
10. Incubate for **5 min** at R.T.
11. Bring it to a final volume of 2.5 mL. So, add 200 μL separation buffer before use of magnet. Mix gently 2–3 times.
12. Place the tube in the magnet without the cap for **2.5 min**.
13. Pour off in one motion with the magnet the solution into a 15 mL tube. Leave the magnet in inverted position (45° angle) for 3 s, and then return to upright position.
14. Count pNK cells* (the NK cells are 5–15% of total PBMC's**) as mentioned above in STEP 14. ****

≅20–45*10^6 pNKs were obtained from 50 mL blood (from total PBMCS**).

** **Donor dependent**
**** **Experimental stop point. The obtained primary NK cells can be seeded to 96 well plates before KIR2DL1 isolation.**

15. Stain with anti- CD56 and anti-CD3 to check purity in the FACS (Fig. 1A)

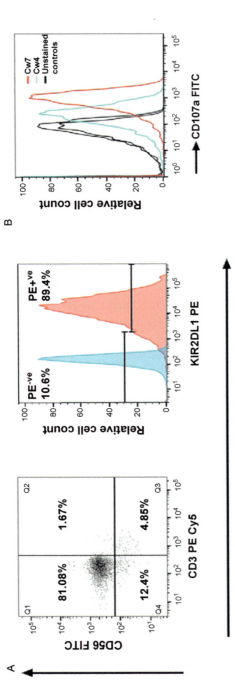

Fig. 1 Determining the purity of pNK isolation with magnetic beads and pNK-KIR2DL1 cells enrichment. (A) pNK were stained for 30 min on ice with anti-CD3-PE-Cy5 and anti-CD56-FITC followed by FACS analysis to check their purity (left panel). Next, pNKs were subjected to KIR2D separation using magnetic beads. Cells were dissociated from the magnetic beads, and the sub-population obtained from the magnetic beads (represent the KIR positive sub-population) and the sub-population obtained from the washes (represent the KIR negative sub-population) was checked for purity by FACS analysis (right panel). The purity of the pNK KIR2D$^+$ population was found to be 89.4%. (B) Isolated pNK KIR2DL1$^+$ cells as detailed above were incubated with 721.221 HLA-Cw4 (inhibition) or 721.221 HLA-Cw7 (activation) for 5 h followed by staining with anti-CD107a for 30 min (degranulation marker) on ice and FACS analysis. Representative histogram showing that the pNK KIR2DL1$^+$ cells exhibited reduced degranulation upon inhibitory interaction (721.221 HLA-Cw4) – blue histogram, vs activating interaction (721.221 HLA-Cw7) – red histogram is shown.

3.2 Isolation for KIR2DL1 positive population

1. Add up to 5 mL separation buffer and centrifuge the cells for **7** min at 1400 rpm (400 g).
2. For each concentration of $<10^7$ cells resuspend in 90 μL separation buffer and transfer to FACS tube (Falcon® Round-Bottom Polystyrene Tubes, 5 mL).
3. Add 10 μL of $F_{c\gamma}R$ blocking antibody and incubate for 5 min at R.T.
4. **Turn off light** in the hood and add: CD158a/h-PE antibody (Miltenyi Biotec 130-092-684) 1:11—10 μL, pipette well and incubate for **10 min** at 4 °C.
5. Wash twice with 5 mL separation buffer and centrifuge for **5** min at 1400 rpm (400 g).
6. Resuspend in 90 μL separation buffer.
7. Add 20 μL of EasySep PE selection Cocktail (positive selection kit). Mix well and incubate at R.T. for **15 min**.
8. Mix the magnetic beads tube with 1 mL pipette and <u>do not</u> pipette up and down more than five times. Add 40 μL magnetic beads into the FACS tube. Mix well and incubate at R.T for **10** min.
9. Bring the cell suspension to a total volume of 2.5 mL (add 2.34 mL of separation buffer). Mix gently 2–3 times with 1 mL pipette.
10. Place the tube <u>without the cap</u> into the magnet and incubate at R.T for **5 min.**
11. Pour off in one motion with the magnet the solution into 50 mL tube. Leave the magnet in inverted position (45° angle) for 3 s, and then return to upright position.
12. Take out the tube from the magnet. Add 2.5 mL separation buffer. Mix the solution gently until it's homogenized. Place the tube back in the magnet and incubate at R.T for **5 min.**
13. Repeat steps 11–12 <u>3 times</u> until you have 7.5 mL (Final Volume) in the collection tube. These are KIR^- cells.
14. Take out the tube from the magnet and add 2 mL separation buffer. Wash the tube carefully (these are the desired cells, primary NK KIR^+).
15. Count the cells*.
    * $\cong$89–95% of pNKs were found to be KIR+ depending on the cell number at starting point.
16. Check purity in the FACS (Fig. 1A).

Take samples from both KIR$^-$ and KIR$^+$ tubes. In addition, stain with anti- CD56 and anti-CD3. (If you didn't check purity for the pervious isolation).

The NK cells obtained are maintained in cell culture according to experimental needs.

Note! Due to the lack of available antibodies capable of distinguishing inhibitory and activating receptors KIR2DL1 and KIR2DLS1 respectively two control group can be used in each experiment to determine specific sub-populations as previously shown by us (Ben-Shmuel et al., 2022; Biber et al., 2021; Sabag et al., 2022). pNK KIR2DL/S1$^+$ were incubate with 721.221 HLA-Cw4 (inhibition) or with 721.221 HLA-Cw7 followed by CD107a stain in order to measure the degranulation ability of pNK KIR2DL/S1$^+$ in both conditions (Fig. 1B).

4. Concluding remarks

NK cells should express at list one inhibitory receptor specific to MHC-I molecule(s) expressed by the organism. NK cells lacking expression of all MHC-I inhibitory receptors or NK cells from MHC-I deficient individuals are incapable of rejecting susceptible, cancerous target cells, are defined as hyporesponsive cells' (Ardolino et al., 2014; Frutoso & Mortier, 2019; Judge, Murphy, & Canter, 2020).

NK isolation and specific KIR population enrichment protocol is highly valuable in advancing our understanding of NK cell biology. This protocol can be applied in various research aspects, depending on the availability of specific antibodies. It allows for the isolation and investigation of any desired KIR, enabling researchers to explore inhibitory and activating interactions associated with specific KIRs. It can be used to isolate any KIR provided the availability of the antibody. The applicability varies from investigating inhibitory and activating interactions of specific KIR's and also to establish standardization of KIR NK cells for potential immunotherapies.

Furthermore, this protocol has the potential to contribute to the standardization of Chimeric Antigen Receptor (CAR) NK cells. CAR NK cells are engineered NK cells expressing receptors that specifically recognize and target cancer cells. By utilizing the established protocol, researchers can isolate and study specific KIRs, which may aid in the development and optimization of CAR NK cells. Understanding the inhibitory and activating properties of KIRs can guide the design and implementation of effective CAR NK cell therapies, enhancing their potential in cancer immunotherapy. By unraveling the intricate mechanisms of NK cell function, we can advance our knowledge and potentially harness the therapeutic potential of NK cells in combating cancer and other NK-related diseases.

Acknowledgments

This research was funded by the Israel Science Foundation (ISF) grant no. 1001/23.

References

Anfossi, N., et al. (2006). Human NK cell education by inhibitory receptors for MHC class I. *Immunity*, *25*, 331–342.

Ardolino, M., et al. (2014). Cytokine therapy reverses NK cell anergy in MHC-deficient tumors. *The Journal of Clinical Investigation*, *124*, 4781–4794.

Ben-Shmuel, A., Biber, G., Sabag, B., & Barda-Saad, M. (2021). Modulation of the intracellular inhibitory checkpoint SHP-1 enhances the antitumor activity of engineered NK cells. *Cellular & Molecular Immunology*, *18*, 1314–1316.

Ben-Shmuel, A., et al. (2022). Inhibition of SHP-1 activity by PKC-θ regulates NK cell activation threshold and cytotoxicity. *eLife*, *11*.

Béziat, V., et al. (2013). Influence of KIR gene copy number on natural killer cell education. *Blood*, *121*, 4703.

Biber, G., et al. (2021). Modulation of intrinsic inhibitory checkpoints using nano-carriers to unleash NK cell activity. *EMBO Molecular Medicine*, *e14073*. https://doi.org/10.15252/EMMM.202114073.

Frutoso, M., & Mortier, E. (2019). NK cell Hyporesponsiveness: More is not always better. *International Journal of Molecular Sciences*, *20*.

Hsu, H. T., et al. (2016). NK cells converge lytic granules to promote cytotoxicity and prevent bystander killing. *The Journal of Cell Biology*, *215*, 875.

Judge, S. J., Murphy, W. J., & Canter, R. J. (2020). Characterizing the dysfunctional NK cell: Assessing the clinical relevance of exhaustion, Anergy, and senescence. *Frontiers in Cellular and Infection Microbiology*, *10*, 49.

Lanier, L. L. (2008). Up on the tightrope: Natural killer cell activation and inhibition. *Nature Immunology*, *9*.

Long, E. O., Sik Kim, H., Liu, D., Peterson, M. E., & Rajagopalan, S. (2013). Controlling natural killer cell responses: integration of signals for activation and inhibition. *Annual Review of Immunology*, *31*, 227–258.

Matalon, O., et al. (2016). Dephosphorylation of the adaptor LAT and phospholipase C – g by SHP-1 inhibits natural killer cell cytotoxicity. *Science Signaling*, *9*, 1–16.

Matalon, O., et al. (2018). Actin retrograde flow controls natural killer cell response by regulating the conformation state of SHP-1. *The EMBO Journal*, *37*, e96264.

Raulet, D. H. (2006). Missing self recognition and self tolerance of natural killer (NK) cells. *Seminars in Immunology*, *18*, 145–150.

Sabag, B., et al. (*2022*). Actin retrograde flow regulated by the Wiskott–Aldrich syndrome protein drives the natural killer cell response. *Cancers*, *14*, 3756.

Vivier, E. (2008). Functions of natural killer cells. *Nature Immunology*, *9*, 503–510.

Yawata, M., et al. (2008). MHC class I–specific inhibitory receptors and their ligands structure diverse human NK-cell repertoires toward a balance of missing self-response. *Blood*, *112*, 2369.

Yu, J., et al. (2007). Hierarchy of the human natural killer cell response is determined by class and quantity of inhibitory receptors for self-HLA-B and HLA-C ligands. *Journal of Immunology*, *179*, 5977–5989.

> CHAPTER TEN

Flow cytometry conjugate formation assay between natural killer cells and their target cells

Gilles Iserentant, Carole Seguin-Devaux, and Jacques Zimmer* ⓘ

Department of Infection and Immunity, Luxembourg Institute of Health, Esch-sur-Alzette, Luxembourg
*Corresponding author: e-mail address: jacques.zimmer@lih.lu

Contents

1. Introduction	214
2. Materials	216
2.1 Equipment	216
2.2 Cell lines	216
2.3 Reagents	216
3. Methods	217
3.1 Culture of effector KHYG-1 NK cell line and target K562/Raji cell lines	217
3.2 Isolation of untouched peripheral blood NK (pNK) cells from cryopreserved PBMC	217
3.3 Cell trace protocol staining of effector and target cells	218
3.4 Effector and target cells co-culture (see Fig. 1 for a schematic outline of the assay)	219
4. Results	219
5. Discussion	225
References	227

Abstract

Before being able to kill other cells, natural killer (NK) cells first have to establish contact with those targets. In case of a predominance of activating signals from the target cell over inhibitory ones, the killing process is initiated. It is possible, with a simple two-color flow cytometry method, to evaluate, for any given effector cell–target cell pair, the number of conjugates between both types of cells. The percentage obtained gives an idea of the amplitude of binding of the NK cells to the targets and might be expected to be indicative of the level of cytotoxicity. Nevertheless, there is no absolute correlation, as the percentages of conjugates are sometimes higher with relatively resistant targets than with the highly sensitive cell line K562. Practically, NK cells and target cells are stained with two differently fluorescent dyes and incubated together at the desired

---

Methods in Cell Biology, Volume 193
ISSN 0091-679X
https://doi.org/10.1016/bs.mcb.2024.02.037

Copyright © 2025 Elsevier Inc.
All rights are reserved, including those
for text and data mining, AI training,
and similar technologies.

213

effector:target ratio (in our example, 1:1) for various periods of time (0, 10, 30 min, etc.) at 37 °C. After the incubation time, the cells are carefully introduced into the flow cytometer, where in principle three populations are distinguished: the single positive, unconjugated effector and target cells, respectively, and the double positive subset, which corresponds to the conjugates between both cell types. We describe here in detail the staining and cell culture protocols and procedures, and give several examples. Thus, the very cytotoxic NK leukemia cell line KHYG-1 *versus* the myeloid leukemia K562 (the "conventional" NK cell target) and the Burkitt lymphoma cell line Raji forms a high number of conjugates. In contrast, purified, non-activated, healthy donor-derived peripheral blood NK cells bind less to the targets, in accordance with their low (K562) or absent (Raji) cytotoxic activity.

1. Introduction

Natural killer (NK) cells are cytotoxic lymphocytes of the innate lymphoid cell (ILC) family that play important roles in the immune defense against various microbes (viruses, bacteria, parasites) as well as against tumor cells. For this purpose, they rely on three main mechanisms (Caligiuri, 2008; Demaria et al., 2019):

(1) natural killing, which consists in the direct (no activation nor immunization needed) lysis of the infected target or the cancer cell,

(2) antibody-dependent cellular cytotoxicity (ADCC), characterized by the crosslinking of the NK cell and the pathogenic cell *via* an antibody (bound to CD16 on the NK cell *via* the Fc part and on a target cell surface structure *via* the Fab part),

(3) abundant cytokine production, important in the early stage of the immune response and able to shape adaptive immunity.

These processes are tightly regulated and governed by the outcome of a balance between messages transmitted by activating receptors on the one hand and by inhibitory receptors on the other hand. Upon target encounter, NK cells form an immunological synapse with the other cell and integrate the activating and inhibitory messages, respectively. The outcome decides about the life or the death of the target cell, the latter occurring when the activating signals overtly predominate (Caligiuri, 2008; Demaria et al., 2019).

Natural killer cells eliminate their targets through two main mechanisms. First, the release of cytotoxic molecules such as perforin, granzymes and, in human, granulysin, contained within intracellular granules and secreted upon NK cell stimulation in the presence of infected cells or tumor cells, induces apoptosis within a few hours. Furthermore, surface molecules like FasL and TRAIL, that interact with their membrane receptors on the target cells, similarly provoke apoptosis, but this process takes longer than the granule-mediated killing (Caligiuri, 2008; Demaria et al., 2019).

Peripheral blood NK cells are the best studied ILC due to their easy accessibility through venous blood draw. However, NK cells have to some extent organ- and tissue-specific features with a different phenotype and functional behavior, so that observations from peripheral blood are not always generalizable (Freud, Mundy-Bosse, Yu, & Caligiuri, 2017). Nevertheless, it is quite convenient to work with these cells, although one has to take into account that the data are not necessarily representative for NK cells from other organs.

Among peripheral blood NK cells, there are classically several populations defined by the level of expression of the adhesion molecule CD56 and of the activating receptor CD16. The major subsets are $CD56^{bright}CD16^-$, mainly producing cytokines and more immature, and the numerically predominant $CD56^{dim}CD16^{bright}$ NK cells, endowed with the ability of natural cytotoxicity and ADCC and more mature. If activated, both subsets are cytotoxic and secrete cytokines (particularly interferon (IFN)-γ) (Cooper, Fehniger, & Caligiuri, 2001; Zimmer, 2020). The additional subpopulations are $CD56^{bright}CD16^{dim}$, $CD56^{dim}CD16^-$ (which corresponds to recently activated cells having transiently lost CD16 expression), $CD56^-CD16^{bright}$, expanded in chronic viral infections but dysfunctional, and, for some authors, $CD56^{dim}CD16^{dim}$ NK cells, displaying a different and relatively immature phenotype compared to their $CD56^{dim}CD16^{bright}$ counterparts (Cooper et al., 2001; Zimmer, 2020).

As the immunological synapse is described in much detail and under various aspects in the three volumes of this book, we will focus on the NK cell–target cell conjugate formation test by flow cytometry, which is an experiment that still can deliver interesting and useful information about the conjugate formation capability of NK cells, a prerequisite for target killing. The precise depicting of the immune synapse is to some extent complicated and not necessarily routinely available in conventional immunology laboratories. However, this very simple assay, which requires only two-color flow cytometry, allows to determine, among a NK cell–target cell mix at the 1:1 ratio (or other ratios if preferred), the percentage of NK cells or target cells that are bound together in conjugates (and have formed immunological synapses, although the latter cannot be appreciated with this method) (Burshtyn & Davidson, 2010).

The principle of the assay relies on the staining of NK cells and target cells with two fluorescent membrane dyes emitting light at different, non-overlapping, wave lengths. Then, the NK cells are incubated with the target cells at, for example, a 1:1 ratio for various time points, like 10 min, 30 min, 1 h, etc. At each time point, when analyzed by the flow cytometer, the percentage of NK cell–target cell conjugates corresponds to the percentage of double positive events among the total events measured. Single positive cells for each of the two dyes correspond to unconjugated NK cells and target cells, respectively. The staining, conjugation and analyzing procedures are described in detail below.

2. Materials

- Polystyrene T-75 (75 cm^2) flasks
- Polystyrene T-25 (25 cm^2) flasks
- Serological pipettes (5 and 10 mL)
- V-bottom-96-well plates
- 15 and 50 mL Falcon™ conical centrifuge tubes
- LS Columns (Miltenyi Biotec, Ref: 130-042-401)

2.1 Equipment
- Laminar flow hood (safety level II)
- Incubator for standard cell culture conditions in a humidified atmosphere (37 °C, 5% CO_2)
- Centrifuge
- Micropipettes, multichannel micropipettes and tips
- Manual separators for magnetic cell isolation
- Flow cytometry instrumentation

2.2 Cell lines
- KHYG-1 NK cell line (DSMZ: ACC725)
- K562 cells (ATCC: CCL-243)
- Raji cells (ATCC: CCL-86)

2.3 Reagents
- Cell culture medium: RPMI1640
- Cell culture medium: IMDM (Iscove's modified Dulbecco's medium)

- Complete medium: RPMI1640 or IMDM supplemented with 10% heat-inactivated Fetal Bovine Serum (FBS), 2 mM L-glutamine, 100 U/mL Potassium Penicillin, 100 μg/mL Streptomycin Sulfate (Pen-Strep)
- Recombinant Human IL-2 Protein (R&D Systems, Ref: 202-IL)
- Human Recombinant Human IL-15 Protein (Stemcell Technologies, Ref: 78031)
- Phosphate Buffered Saline (PBS)
- CellTrace™ Violet Cell Proliferation kit (ThermoFisher, Ref: C34557)
- CellTrace™ CFSE Cell Proliferation kit (ThermoFisher, Ref: C34554)
- NK Cell Isolation Kit, human (Miltenyi Biotec, Ref: 130-092-657)
- MACS Buffer: dilute MACS® Bovine Serum Albumin (BSA) Stock Solution (Ref: 130-091-376) 1:20 with autoMACS® Rinsing Solution (Ref: 130-091-222), both from Miltenyi Biotec

3. Methods

3.1 Culture of effector KHYG-1 NK cell line and target K562/Raji cell lines

K562 cells are cultured in T-75 flasks with 15 mL complete IMDM medium at a starting density of 1×10^5 viable cells/mL. Subculture by adding fresh medium every 2–3 days at a density of 1×10^6 viable cells/mL.

Raji cells are cultured in the same way but in complete RPMI1640 medium at a starting density of 4×10^5 viable cells/mL and subcultured at a maximum of 3×10^6 viable cells/mL every 2–3 days.

KHYG-1 cells are cultured in T-25 flasks with 5 mL of complete RPMI1640 medium supplemented with 100 IU/mL of recombinant IL-2. Cells are maintained at a density of approximately 1×10^6 viable cells/mL and subcultured accordingly every 2–3 days.

3.2 Isolation of untouched peripheral blood NK (pNK) cells from cryopreserved PBMC

Cryopreserved PBMC are gently thawed in a 37 °C water bath. PBMC are washed in 50 mL of PBS and centrifuged at $300 \times g$ for 10 min at room temperature. After washing, the PBMC are resuspended in 5 mL of pre-heated complete RPMI culture medium. After cell count, untouched pNK cells are isolated from PBMC using the NK Cell Isolation kit from Miltenyi Biotec as follows:
- Determine exact cell number
- Transfer total cells in a 15 mL Falcon™ conical centrifuge tube

- Centrifuge cell suspension at $300 \times g$ for 10 min and aspirate supernatant
- Resuspend cell pellet in 40 μL of cold MACS buffer per 10^7 total cells
- Add 10 μL of NK Cell Biotin-Antibody Cocktail per 10^7 total cells
- Mix gently and incubate at 4 °C for 10 min
- Add 30 μL of cold MACS buffer per 10^7 total cells
- Add 20 μL of NK Cell MicroBead Cocktail per 10^7 total cells
- Mix gently and incubate for 15 min at 4 °C
- Wash cells by adding 2 mL of cold MACS buffer per 10^7 total cells
- Centrifuge at $300 \times g$ for 10 min and aspirate supernatant
- Resuspend up to 10^8 cells in 500 μL of cold MACS buffer
- Place the LS MACS column in the MACS Separator
- Wash column with 3 mL of cold MACS buffer
- Apply cell suspension onto the column
- Perform three washing steps with 3 mL of cold MACS buffer
- Collect total effluent, as this fraction represents the enriched NK cells
- Centrifuge at $300 \times g$ for 10 min and aspirate supernatant

After isolation, untouched pNK cells are washed with 15 mL of PBS and resuspended in 2 mL of PBS for counting. Adjust cell concentration at 1×10^6 cells/mL and proceed with CellTrace™ staining.

3.3 Cell trace protocol staining of effector and target cells

Freshly cultured K562 and Raji cells are washed with 15 mL PBS and adjusted to a cell concentration of 10^6 cells/mL after counting. Proceed with CellTrace™ staining of the K562 and Raji target cells with the CellTrace™ Violet Proliferation kit. Prepare the CellTrace Violet staining stock solution by adding 20 μL of DMSO into the vial. Add 1 μL of stock solution per 10^6 cells/mL for a 5 μM staining solution and incubate for 20 min at 37 °C. Centrifuge the cells at $300 \times g$ for 10 min and perform two washing steps with 10 mL of pre-heated PBS. Resuspend cells in their respective pre-heated complete culture medium after the second wash, perform cell count and adjust cell concentration at 10^6 cells/mL.

Untouched pNK cells or KHYG-1 are stained with CellTrace™ CFSE Proliferation kit. Staining of untouched pNK cells should be performed straight after isolation for the non-activated conjugate test. For a conjugate test with shortly activated untouched NK cells, 2×10^6 NK cells are incubated for 24 h in a 24-well plate in 2 mL of complete RPMI medium supplemented with 100 IU/mL of human recombinant IL-2 and 10 ng/mL of human recombinant IL-15. Untouched pNK cells are stained with CellTrace™ CFSE Proliferation kit prior to the conjugate test. Prepare the CellTrace™ CFSE staining stock solution by adding $18\,\mu L$ of DMSO into the vial. Add $0.9\,\mu L$ of stock solution per 10^6 cells/mL for a $5\,\mu M$ staining solution and incubate for 20 min at $37\,^{\circ}C$. Centrifuge the cells at $300 \times g$ for 10 min and perform two washing steps with 10 mL of pre-heated PBS. Resuspend cells in pre-heated complete RPMI culture medium after the second wash, perform cell count and adjust cell concentration at 10^6 cells/mL.

3.4 Effector and target cells co-culture (see Fig. 1 for a schematic outline of the assay)

After CellTrace™ staining, cell counts are assessed and cells are coincubated at E:T ratio 1:1 for 2 h, 1 h and 30 min, respectively, at $37\,^{\circ}C$ in V-bottom 96 well plates. NK effector cells (KHYG-1 and untouched pNK) and target cells (Raji and K562) are separately resuspended at a final cell concentration of 10^6 cells/mL. At each time point, $100\,\mu L$ of effector cells are pooled together with $100\,\mu L$ of target cells in a 96 well V-bottom plate in triplicate. Pooled cells are then centrifuged at $200 \times g$ for 1 min to enhance contact between effectors and targets. After pooling cells together for the last time point T0, cells are centrifuged as explained before and supernatant is removed gently. As a control, effector and target cells are also incubated alone as a conjugate negative control. Resuspend the cell pellet with $100\,\mu L$ of 1% PFA solution very slowly to not break aggregates. Acquisition is done with a conventional flow cytometer and analyzed with its analysis software. Using side scatter (SSC) and forward scatter (FSC) parameters, the integrity of the cells is assessed and dead cells and debris, which have a reduced light scattering, are excluded. Create a gate called "Cells" including all viable cells. Plot data (FSC) *versus* time "Stability" should be done next to assess a stable flow rate and avoid clogs or pressure problems which could result in data loss. An intersection with gates "Cells" and "Stability" should be created for further analysis. This final intersect gate is used for plotting CellTrace™ CFSE against CellTrace™ Violet to visualize conjugates between both effector and target cells. Quadrant gating should be placed with help of the different cells' negative controls to ensure specific conjugate gating. This is illustrated in Fig. 2.

4. Results

We performed conjugate formation assays between K562 and Raji target cells and, as effector cells, either the NK cell leukemia line

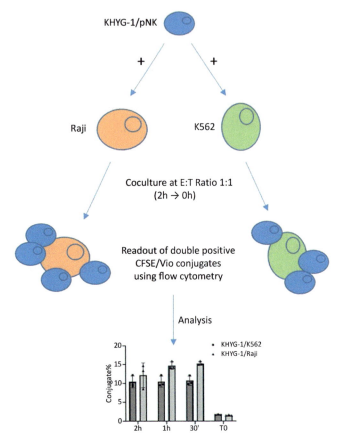

Fig. 1 Schematic representation of the assay setup. Target cells (K562 and Raji) and effector cells (KHYG-1 and purified NK cells isolated from peripheral blood) are stained with two different fluorescent dyes and cultured together for various time points. At the end of the incubation period, the number of conjugates, corresponding to the percentage of events in the double positive gate, is assessed by flow cytometry.

KHYG-1, known to be highly cytotoxic *in vitro* (Hu, Kosaka, Marcus, Rashedi, & Keating, 2019), or untouched peripheral blood NK cells from healthy donors that were used *ex vivo* or pre-activated *in vitro* with the cytokines IL-2 and IL-15.

KHYG-1 cells formed a relatively high number of conjugates with both target cell lines (Fig. 3). The percentage of cells in conjugates at T0 was either very low or high depending on the exact procedure. Indeed, when the samples were not fixed at T0 (Fig. 4), the number of conjugates was very low (<5%) with both target cell lines, and then increased with time of

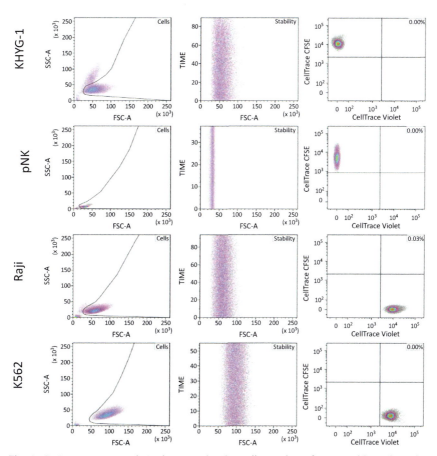

Fig. 2 Gating strategy and single controls. The cells used are first gated based on the FSC–SSC (size and granularity, respectively) parameters (left column). Then, stability of the measurement is assessed in the FSC *versus* Time plots (middle column). The different widths in these plots reflect the various FSC's of the cells. Finally, the right column represents single stained samples (CellTrace Violet *versus* CellTrace CFSE) of the four cell types of the assay. The corresponding tubes might also serve for compensation purposes.

incubation (Fig. 4). In contrast, in the case of fixation with 1% PFA, the conjugate percentages at T0 reached up to 40% with Raji and 10% with K562 (Fig. 3). This issue had already been commented by Burshtyn and Davidson (2010), who attributed it to a binding that occurs during the centrifugation of the samples.

Peripheral, resting *ex vivo* NK cells also formed conjugates with both K562 and Raji when the samples were fixed (Fig. 5). However, the

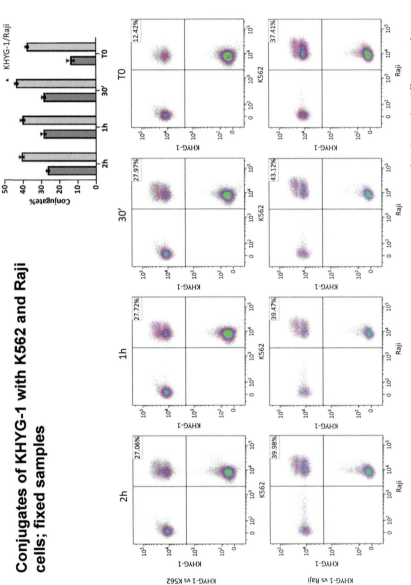

Fig. 3 Conjugate formation between the NK cell line KHYG-1 and the tumor targets K562 and Raji; fixed samples. Effector and target cells were cultured together for the indicated periods of time, fixed with PFA, and conjugates measured by flow cytometry, as described in Sections 2 and 3. The fixation is likely responsible for the high number of conjugates at T0 (Burshtyn & Davidson, 2010). PFA, paraformaldehyde 1%.

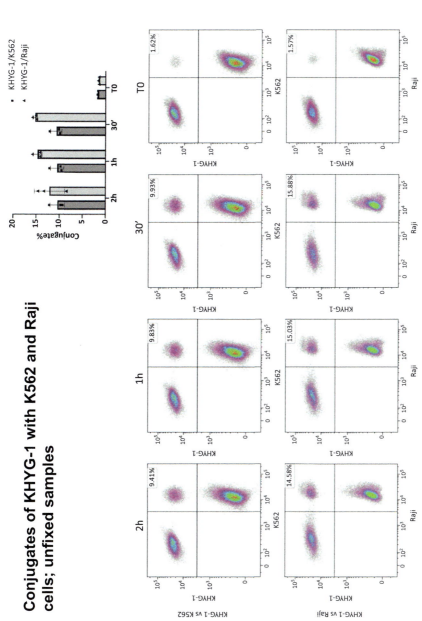

Fig. 4 Conjugate formation between the NK cell line KHYG-1 and the tumor targets K562 and Raji; unfixed samples. Effector and target cells were cultured together for the indicated periods of time with no fixation, and conjugates measured by flow cytometry, as described in Sections 2 and 3.

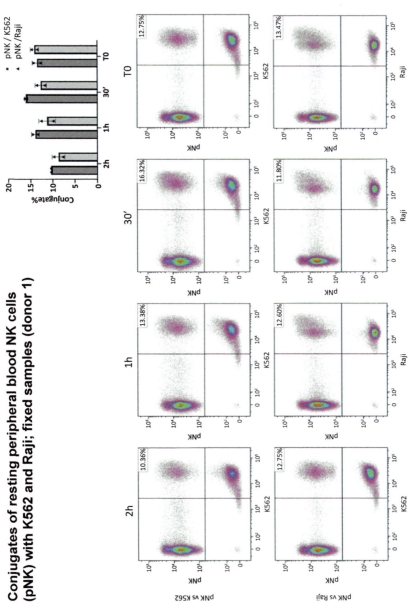

Fig. 5 Conjugate formation between purified resting peripheral blood NK cells and the tumor targets K562 and Raji; fixed samples (donor 1). Conjugate formation between isolated peripheral blood NK cells used immediately after the purification procedure, and the targets K562 and Raji on fixed samples. Culture conditions and readout (flow cytometry) as detailed in Sections 2 and 3.

percentages of double positive events were much lower in the absence of fixation (Fig. 6). This might not be surprising, as Raji is known to be resistant to resting NK cells (Zimmer et al., 1998) and that in a conjugate assay of the NK cell line NK92 *versus* the cervix carcinoma HeLa, the percentage of conjugates peaked at less than 10% after 60 min of co-incubation (Jarahian, Watzl, Issa, Altevogt, & Momburg, 2007). Overall, we observed in some samples, a leakage of the fluorescent dyes out of the stained cells. This has been considered previously as the result of dissociated conjugates occurring after longer incubation periods (Burshtyn & Davidson, 2010), as is the case here. Alternatively, one might consider a trogocytosis phenomenon, well described in NK cells (Campos-Mora et al., 2023; Zimmer, Ioannidis, & Held, 2001).

5. Discussion

NK cell–target cell conjugate formation assays have been used in fundamental research for various purposes. For example, Back et al. studied with this method some characteristics of the *cis-trans* model of NK cell education (Back, Chalifour, Scarpellino, & Held, 2007). Another application of conjugate formation assays is to check if human NK cells have an adhesion deficit to potential target cells in the context of inborn errors of immunity. Jarahian et al. demonstrated, partly based on conjugate formation assays, that the overexpression of the neural cell adhesion molecule NCAM (CD56) by tumor cells inhibits the cytotoxic activity of NK cells (Jarahian et al., 2007). The group of Carsten Watzl described that statins, very frequently used as a treatment of hypercholesterolemia, are able to inhibit NK cell-mediated cytotoxic activity through an influence on the immune synapse-governing LFA-1 (CD11a/CD18) integrin, reducing the number of conjugates between NK cells and tumor targets (Raemer, Kohl, & Watzl, 2009). *In vitro*, the mucin MUC-16, which carries the epitopes of the tumor marker CA125, inhibits conjugate formation between human NK cells and ovarian cancer cell lines (Felder et al., 2019). In contrast, the natural compound piperlongumine increases the number of conjugates between NK cells (cell line YTS) and target cells (various solid tumor lines) (Afolabi, Bi, Chen, & Wan, 2021). The same holds true for the NK cell line NKL and the B lymphoblastoid target 721.221 in the presence of 8-azaguanine (Kim et al., 2019). Witkowski et al. showed that in severe COVID-19 patients, the percentage of NK cells in conjugates with K562 is significantly lower than in healthy control donors (Witkowski et al., 2021). Several further

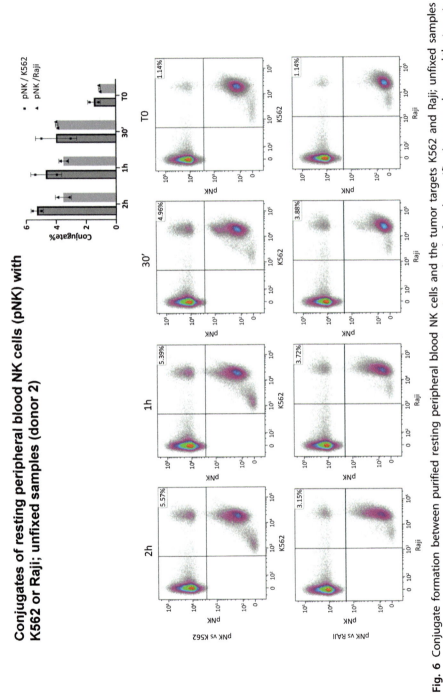

Fig. 6 Conjugate formation between purified resting peripheral blood NK cells and the tumor targets K562 and Raji; unfixed samples (donor 2). Conjugate formation between isolated peripheral blood NK cells used immediately after the purification procedure, and the targets K562 and Raji on unfixed samples. Culture conditions and readout (flow cytometry) as detailed in Sections 2 and 3.

groups have made use of the method, see for example references (Pariani et al., 2023; Wu et al., 2018; Zhang et al., 2016).

Although we describe here the conjugate formation assay with human NK cells and human cancer cell lines as targets, the same method might be applied to other species, such as the mouse and the rat. In this situation however, it is important to use appropriate effectors and targets, like the murine lymphoma cell lines YAC-1, RMA and RMA-S when investigating mouse NK cells. We confirmed here a higher number of conjugates when the cells were fixed as compared to unfixed samples, and recommend therefore not to fix the aliquots if possible.

References

Afolabi, L. O., Bi, J., Chen, L., & Wan, X. (2021). A natural product, piperlongumine (PL), increases tumor cells sensitivity to NK cell killing. *International Immunopharmacology, 96*, 107658. https://doi.org/10.1016/j.intimp.2021.107658.

Back, J., Chalifour, A., Scarpellino, L., & Held, W. (2007). Stable masking by H-2D^d cis ligand limits Ly49A relocalization to the site of NK cell/target cell contact. *Proceedings of the National Academy of Sciences of the United States of America, 104*, 3978–3983. https://doi.org/10.1073/pnas.0607418104.

Burshtyn, D. N., & Davidson, C. (2010). Natural killer cell conjugate assay using two-color flow cytometry. In K. S. Campbell (Ed.), *Methods in molecular biology: Vol. 612. Natural killer cell protocols* (pp. 89–96). https://doi.org/10.1007/978-1-60761-362-6_7.

Caligiuri, M. A. (2008). Human natural killer cells. *Blood, 112*, 461–469. https://doi.org/10.1182/blood-2007-09-077438.

Campos-Mora, M., Jacot, W., Garcin, G., Depondt, M. L., Constantinides, M., Alexia, C., et al. (2023). NK cells in peripheral blood carry trogocytosed tumor antigens from solid cancer cells. *Frontiers in Immunology, 14*, 1199594. https://doi.org/10.3389/fimmu.2023.1199594.

Cooper, M. A., Fehniger, T. A., & Caligiuri, M. A. (2001). The biology of human natural killer-cell subsets. *Trends in Immunology, 22*, 633–640. https://doi.org/10.1016/s1471-4906(01)02060-9.

Demaria, O., Cornen, S., Daëron, M., Morel, Y., Medzhitov, R., & Vivier, E. (2019). Harnessing innate immunity in cancer therapy. *Nature, 574*, 45–56. https://doi.org/10.1038/s41586-019-1593-5.

Felder, M., Kapur, A., Rhakmilevich, A. L., Qu, X., Sondel, P. M., Gillies, S. D., et al. (2019). MUC16 suppresses human and murine innate immune responses. *Gynecologic Oncology, 152*, 618–628. https://doi.org/10.1016/j.ygyno.2018.12.023.

Freud, A. G., Mundy-Bosse, B. L., Yu, J., & Caligiuri, M. A. (2017). The broad spectrum of human natural killer cell diversity. *Immunity, 47*, 820–833. https://doi.org/10.1016/j.immuni.2017.10.008.

Hu, C. H. D., Kosaka, Y., Marcus, P., Rashedi, I., & Keating, A. (2019). Differential immunomodulatory effects of human bone marrow-derived mesenchymal stromal cells on natural killer cells. *Stem Cells and Development, 28*, 933–943. https://doi.org/10.1089/scd.2019.0059.

Jarahian, M., Watzl, C., Issa, Y., Altevogt, P., & Momburg, F. (2007). Blockade of natural killer cell-mediated lysis by NCAM140 expressed on tumor cells. *International Journal of Cancer, 120*, 2625–2634. https://doi.org/10.1002/ijc.22579.

Kim, N., Choi, J. W., Song, A. Y., Choi, W. S., Park, H. R., Park, S., et al. (2019). Direct potentiation of NK cell cytotoxicity by 8-azaguanine with potential antineoplastic activity. *International Immunopharmacology*, *67*, 152–159. https://doi.org/10.1016/j.intimp.2018.12.020.

Pariani, A. P., Almada, E., Hidalgo, F., Borini-Etichetti, C., Vena, R., Marín, L., et al. (2023). Identification of a novel mechanism for LFA-1 organization during NK cytolytic response. *Journal of Cellular Physiology*, *238*, 227–241. https://doi.org/10.1002/jcp.30921.

Raemer, P. C., Kohl, K., & Watzl, C. (2009). Statins inhibit NK-cell cytotoxicity by interfering with LFA-1-mediated conjugate formation. *European Journal of Immunology*, *39*, 1456–1465. https://doi.org/10.1002/eji.200838863.

Witkowski, M., Tizian, C., Ferreira-Gomes, M., Niemeyer, D., Jones, T. C., Heinrich, F., et al. (2021). Untimely TGFβ responses in COVID-19 limit antiviral functions of NK cells. *Nature*, *600*, 295–301. https://doi.org/10.1038/s41586-021-04142-6.

Wu, H., Bi, J., Wu, G., Zheng, C., Lu, Z., Cui, L., et al. (2018). Impaired cytolytic activity of asthma-associated natural killer cells is linked to dysregulated transcriptional program in energy metabolism. *Molecular Immunology*, *101*, 514–520. https://doi.org/10.1016/j.molimm.2018.08.015.

Zhang, M., Bracaglia, C., Prencipe, G., Bemrich-Stolz, C. J., Beukelman, T., Dimmitt, R. A., et al. (2016). A heterozygous *RAB27A* mutation associated with delayed cytolytic granule polarization and hemophagocytic lymphohistiocytosis. *Journal of Immunology*, *196*, 2492–2503. https://doi.org/10.4049/jimmunol.1501284.

Zimmer, J. (2020). CD56^{dim}CD16^{dim} natural killer (NK) cells: The forgotten population. *HemaSphere*, *4*, e348. https://doi.org/10.1097/HS9.0000000000000348.

Zimmer, J., Donato, L., Hanau, D., Cazenave, J. P., Tongio, M. M., Moretta, A., et al. (1998). Activity and phenotype of natural killer cells in peptide transporter (TAP)-deficient patients (type I bare lymphocyte syndrome). *The Journal of Experimental Medicine*, *187*, 117–122. https://doi.org/10.1084/jem.187.1.117.

Zimmer, J., Ioannidis, V., & Held, W. (2001). H-2D ligand expression by Ly49A+ natural killer (NK) cells precludes ligand uptake from environmental cells: Implications for NK cell function. *The Journal of Experimental Medicine*, *194*, 1531–1539. https://doi.org/10.1084/jem.194.10.1531.

CHAPTER ELEVEN

Quantitative analysis of the B cell immune synapse using imaging techniques

Oreste Corrales Vázquez[†], Teemly Contreras[†], Martina Alamo Rollandi, Felipe Del Valle Batalla, and Maria-Isabel Yuseff*

Laboratory of Immune Cell Biology, Department of Cellular and Molecular Biology, Pontificia Universidad Católica de Chile, Santiago, Chile
*Corresponding author: e-mail address: myuseff@bio.puc.cl

Contents

1. Introduction — 230
2. Materials and methods — 230
3. Analysis of B cell activation with immobilized antigens — 234
   3.1 Image and microscopy analysis — 235
4. Analysis of the distribution of organelles in B cells activated with antigen-coated beads — 236
5. Characterization of the immune synapse organization of B cells activated on antigen-coated surfaces — 244
6. Discussion and relevance — 250
7. Notes — 250
References — 252

Abstract

This chapter presents a series of quantitative analyses that can be used to study the formation of the immune synapse (IS) in B cells. The methods described are automated, consistent, and compatible with open-source platforms. The IS is a crucial structure involved in B cell activation and function, and the spatiotemporal organization of this structure is analyzed to provide a better understanding of its mechanisms. The analyses presented here can be applied to other immune cells and are accessible to researchers of diverse fields. In addition, the raw data derived from the results can be further explored to perform quantitative measurements of protein recruitment and tracking of intracellular vesicles. These techniques have the potential to enhance not only our understanding of the IS in B cells but also in other cell models.

[†] Equal contribution.

1. Introduction

B cells are a crucial part of the immune response as they can differentiate into plasma cells that produce and secrete antibodies and become memory B cells that provide long-term protection against new and recurrent infections (Batista & Harwood, 2009). To fulfill these functions, naïve B cells must first be activated in lymphoid tissues upon interaction with antigen-presenting cells (APCs) such as macrophages or dendritic cells. This initial interaction with an APC is known as the immune synapse (IS), a highly organized structure formed between immune cells (Yuseff, Pierobon, Reversat, & Lennon-Duménil, 2013).

The formation of the IS involves a complex and highly regulated series of events comprising cytoskeleton rearrangement, actin remodeling, lysosome recruitment, MTOC polarization, and associated functions such as antigen extraction and changes in cell shape during the IS formation. Actin and microtubule cytoskeletons are remodeled to focus vesicle trafficking important for antigen extraction and processing, as well as to provide a platform for signaling molecules that regulate B cell activation (Yuseff et al., 2011).

To quantify and understand the mechanisms underlying the formation of the B cell IS, imaging techniques are essential. Despite recent advances in techniques and computational tools, the accurate analysis of IS formation remains a challenging task. This is due to variations in the quality and consistency of sample preparation, the heterogeneity of individual cells, and the complexity of the data obtained from imaging analysis. Nevertheless, these challenges can provide important insights and opportunities for the development of improved strategies for data analysis.

Here, we present a series of quantitative analyses of the B cell IS formation using automated and consistent workflows compatible with open-source platforms. The following methods are recommended to minimize bias and ensure reproducibility in data analysis. The use of these analyses can be extrapolated to other immune cells, such as T cells, dendritic cells, macrophages, and other cell models that exhibit polarized phenotypes under different biological contexts.

2. Materials and methods

1. Cells and culture

The mouse B lymphoma A20 cell line was used for experiments that serve as an example. These cells have the phenotype of quiescent mature B-cells expressing surface IgG2a as a BCR (Lankar et al., 1998).

1.1. Culture and experimental medium (CLICK): RPMI 1640 supplemented with 10 mM glutamine, 100 U/mL penicillin, 100 μg/mL streptomycin, 50 μM 2-mercaptoethanol (2-ME), 5 mM sodium pyruvate, and 10% or 5% fetal bovine serum (FBS) for cell culture or activation experiments, respectively.

1.2. 1× phosphate-buffered saline (PBS) sterile.

2. Antibodies and reagents

For additional information about materials **see Note 1**.

3. Preparation of antigen-coated beads

3.1. To activate B cells, use NH_2-beads covalently coated with antigen (antigen-coated beads), which are prepared using 50 μL (~20 * 10^6 beads) of 3 μm NH_2-beads and activating (BCR-ligand$^+$) or non-activating (BCR-ligand$^-$) antigens.

3.2. For A20 B cells, which are an IgG$^+$ FcγR-defective B cell line with the phenotype of quiescent mature B-cells, use the anti-IgG-F(ab′)2 fragment as BCR-ligand$^+$ and anti-IgM-F(ab′)2 or bovine serum albumin (BSA) as BCR-ligand$^-$. See Note 2.

3.3. To prepare antigen-coated beads, place the 50 μL of beads in a low-binding protein microcentrifuge tube containing 1 mL of 1× PBS to maximize the beads recovery during the sample manipulation, and centrifuge at $16,000 \times g$ for 5 min to wash the beads. Discard the supernatant.

3.4. Resuspend the beads with 500 μL of 8% glutaraldehyde to activate the NH_2 groups and rotate for 4 h at room temperature (RT), protected from light.

CAUTION: The glutaraldehyde stock solution should only be used in a chemical fume hood. Follow the instructions on the material safety data sheet (MSDS).

3.5. Centrifuge beads at $16,000 \times g$ for 5 min, remove the glutaraldehyde and wash the beads three times with 1 mL of 1× PBS

CAUTION: Glutaraldehyde solution should be discarded as hazardous chemical waste.

3.6. Resuspend the activated beads in 100 μL of 1× PBS. The sample can be divided into two low-protein binding microcentrifuge tubes: 50 μL for the BCR-ligand$^+$ and 50 μL for BCR-ligand$^-$.

3.7. To prepare the antigen solution use two 2 mL low-protein binding microcentrifuge tubes containing 100 μg/mL of antigen solution in 150 μL of PBS: one tube with BCR-ligand$^+$ and the other with BCR-ligand$^-$.

3.8. Add 50 μL of activated beads solution to each tube containing 150 μL of antigen solution, vortex, and rotate overnight at 4 °C.

3.9. Add $500\,\mu L$ of $10\,mg/mL$ BSA (in PBS $1\times$) to block the remaining reactive NH_2 groups on the beads and rotate for $1\,h$ at $4\,°C$.

3.10. Centrifuge the beads at $16,000 \times g$ for $5\,min$ at $4\,°C$ and remove the supernatant. Wash the beads with cold $1\times$ PBS three times.

3.11. Resuspend activated beads in $40\,\mu L$ of $1\times$ PBS.

3.12. To determine the final concentration of the antigen-coated beads, dilute a small volume of beads ($\sim 1\,\mu L$) in $1\times$ PBS (1:200) and count them using a hemocytometer. Then store at $4\,°C$ until use. **See Note 3**.

4. Preparation of antigen-coated coverslips

4.1 Prepare the antigen solution: $1\times$ PBS containing $10\,\mu g/mL$ BCR-ligand$^+$. **See Note 4**.

4.2 Place the 12- or 10-mm coverslip onto a 24-well plate lid covered with parafilm, add $40\,\mu L$ of antigen solution onto each coverslip, and incubate in a humid chamber at $4\,°C$ overnight. Seal the plate to avoid evaporation of antigen solution.

4.3 Wash the coverslips gently with $100\,\mu L$ of $1\times$ PBS using Pasteur pipettes before continuing to next steps. To prevent the coverslips from drying out, store them in $1\times$ PBS until use.

5. Preparation of poly-L-lysine coverslips

5.1 Prepare $40\,mL$ of 0.01% w/v of poly-L-lysine solution. Use a $50\,mL$ tube to immerse $10\,mm$ coverslips into the solution and rotate overnight at RT. **See Note 4**.

5.2 Wash the poly-L-lysine coverslips (PLL-coverslips) with $1\times$ PBS and leave to air dry on a 24-well plate lid covered with filter paper.

6. B cell activation with antigen-coated beads

6.1. Dilute B cells to a final concentration of $1.5 * 10^6$ cells/mL in CLICK medium with 5% FBS.

6.2. Add 150,000 of antigen-coated beads to $100\,\mu L$ of B cells (150,000) in $0.6\,mL$ tubes to obtain a ratio of 1:1, respectively. Mix gently using a vortex and seed the $100\,\mu L$ onto the PLL-coverslip. Incubate for different time points in a cell culture incubator ($37\,°C$ and 5% CO_2). The typical activating time points are 0, 30, 60, and $120\,min$. For time $0\,min$, place the PLL-coverslips into the 24-well plate lid on ice. Add the antigen-bead mixture to the cells and incubate on ice for 5 min. **See Note 5**.

6.3. After each time point, carefully aspirate the media off each coverslip and add $100\,\mu L$ of cold $1\times$ PBS ($1–3\,min$) to stop the activation. Continue with the immunofluorescence protocol. See step 8 for immunofluorescence.

7. B cell activation on antigen-coated surface
   7.1. Dilute B cells to a final concentration of $1.0 * 10^6$ cells/mL in CLICK medium with 5% FBS.
   7.2. Add $80\,\mu L$ of B cells (80,000) onto an antigen-coated coverslip (antigen-coverslip) and activate for different time points in a cell incubator at $37\,°C/5\%$ CO_2. The typical activating time points are 0, 15, 30 and 60 min. For time 0, use PLL-coverslip and add the cells. Incubate for 5 min.
   7.3. After each time point, carefully aspirate the media off each coverslip and add $100\,\mu L$ of cold $1\times$ PBS (1–3 min) to stop the activation. Continue with the immunofluorescence protocol. See Section 8.
8. Immunofluorescence
   8.1. Remove the $1\times$ PBS and proceed with the fixation of each coverslip. **See Note 6**.
   8.2. Add $50\,\mu L$ of cold 3% PFA in PBS $1\times$ and incubate for 10 min at (RT).
   8.3. **CAUTION:** PFA is toxic. Please read the MSDS before working with this chemical. PFA solutions should only be prepared under a chemical fume hood wearing gloves and safety glasses. PFA solution should be discarded as hazardous chemical waste.
   8.4. Wash three times with $1\times$ PBS as explained in 6.3. **See Note 7.**
   8.5. Remove $1\times$ PBS and add $50\,\mu L$ of blocking buffer (2% BSA and 0.3 M glycine in $1\times$ PBS) onto each coverslip and incubate at RT for 10 min. Wash off the blocking buffer with $1\times$ PBS.
   8.6. Prepare the antibodies dilution in permeabilization buffer (0.2% BSA and 0.05% saponin in $1\times$ PBS). Place $40\,\mu L$ drops of primary antibody solution on a 24-well plate lid or a plastic petri dish lid covered with parafilm. Place the coverslips with fixed cells over each drop and incubate in a humid chamber at $4\,°C$ overnight. Seal the plate to avoid evaporation of antigen solution. **See Note 7 and 8**.
   8.7. Wash the coverslips three times with permeabilization buffer.
   8.8. Dilute the secondary antibodies or dyes in permeabilization buffer, using $40\,\mu L$ per coverslip. Place the coverslips over drops of secondary antibody dilutions and incubate for 1 h at RT in dark and humid chambers.
   8.9. Wash the coverslips twice with permeabilization buffer and once with $1\times$ PBS.
   8.10. Remove the PBS solution from the coverslips.

8.11. Add 5 μL of mounting reagent to a microscope slide. Mount the coverslips onto the slide with the cell side facing down. Allow the slides to dry for 30 min at 37 °C or RT overnight protected from light. **See Note 9**.

8.12. Acquire fluorescence images on a confocal or widefield microscope with a 60× or 100× objective according to sample and cell size. Images acquired for bead assays were obtained in a Nikon Ti Eclipse inverted microscope with 60×/1.45NA oil objective. For spreading assays, a Zeiss LSM880 Confocal microscope (Airyscan detector) with 63×/1.4NA oil immersion lens was used. **See Note 10**.

3. Analysis of B cell activation with immobilized antigens

As mentioned above, cytoskeleton rearrangement, centrosome, and lysosome polarization to the antigen contact site are prerequisites and key hallmarks of immune synapse formation. The following protocols detail B cell activation using antigen-coated beads and surfaces with BCR^+ or BCR^- ligands that emulate an APC. The selection of either method of activation depends on the user needs and experimental design.

The activation of B cells with antigen-coated beads is useful to evaluate and polarization of intracellular structures (e.g., cytoskeleton, organelles) towards a particular region of the cell, in this case, a surface mimicking an Antigen presenting cell. One of the limitations of this technique is the number of beads that randomly become in contact with cells under the experimental conditions. The user should be aware that finding cells in contact with beads at shorter times of activation could be challenging. On the other hand, antigen-coated surfaces, such as coverslips are suitable for analyzing organelle distribution and dynamics at the IS in 2D and 3D, and also allows the user to employ high-resolution techniques, such as TIRF microscopy, to explore features of the synaptic plane. A caveat of this method is the limited control of the amount of ligand adsorbed and the uneven distribution on the surface. Reversat et al. (2015) and Yuseff et al. (2011) illustrate examples using Antigen-coated Beads and Antigen-coated coverslips, respectively.

We will explain the image analysis to study the recruitment of different organelles during B cell IS formation in B cells. These methods can be extrapolated to other cell models that present polarized phenotypes or differential recruitment of organelles to discrete cell regions.

3.1 Image and microscopy analysis

It is important for the user to consider the different alternatives to perform quantitative imaging analysis when studying structural features of the immune synapse. For observations of B cell receptor distribution the most suitable approach is to use a high numerical aperture objective and TIRF illumination, which due to the short penetration depth of the evanescent field (~100 nm) improves signal-to-noise ratio for observing movement of molecules at the cell membrane. If not available, confocal microscopy can provide sufficient resolution to perform 3D imaging and determine colocalization of labels, organelle distribution and particle segmentation. For other experiments, a widefield illumination microscope is useful to quantify organelle distribution and other measurements that we show in this chapter.

After microscopy and image acquisition, we advise users to save raw images in .tif format at the highest bitrate. It is also convenient to use an output as a single file that preserves Z-slices and channels, such as hyperstacks. Fiji has a built-in macro .tif converter available in the "Process" → "Batch" → "Convert…" menu that works for these purposes. For convenient file handling and posterior use, we strongly suggest that files be named correctly with channel information (e.g., staining) and assigned pseudocolor. We also recommended adding a brief description of the experiment in a text file inside the image folder.

The following algorithms are described for Fiji—ImageJ software (Arena et al., 2017), which have been tested and work in versions 1.52-1.53t of said software. We recommend downloading the Bio-formats plugin for importing image files. No other plugins are needed for running the macros used in this chapter. However, this can be performed using equivalent software such as Napari or others if instructions are adjusted to those working environments. For R scripts, all codes were written and tested in R 4.2.2, packages Tidyverse and Readr are needed.

The mentioned ImageJ-compatible macros and R scripts are available to download at https://github.com/YuseffLab/Image-Analysis. All codes are commented to be self-explanatory. If the user needs more details on how to use the R language, several open source projects are available for learning the fundamental concepts for use of this programming language (https://swirlstats.com/students.html, https://r4ds.had.co.nz/).

As we have done for this chapter, for statistical analysis, we recommend using 3 individual biological replicates with at least 20 cells per experiment. It is critical to check for normality of data and remove outliers when identified by statistical tests.

Finally, for all the following Fiji instructions the user should configure the "Set measurements" menu. This can be found in the "Analyze" tab. Select all measurements except "Limit to threshold," "Invert Y coordinates," "Add to overlay," "Scientific notation" and "NaN empty cells." This will provide all of the information needed for the analyses.

4. Analysis of the distribution of organelles in B cells activated with antigen-coated beads

The next protocol details the steps to quantify the polarization of cell components to the IS. In this case, we define an arbitrary value, the polarity index, as a measure of proximity to the IS. The index ranges between −1 (anti-polarized) and 1 (fully polarized, object on the bead), as previously presented by Reversat et al. (2015).

1. Estimate the polarity index for the centrosome: **See Note 11**

We use this algorithm to analyze the polarity of organelles that display a discrete point. In this case, we have labeled the centrosome with α-tubulin (see Table 1 for antibodies and reagents).

Table 1 Antibodies and reagents.

Primary/ secondary antibody	Host	Reactivity	Catalog ID	Observations
Anti-α-tubulin	Rat	Mouse	ab6160, Abcam	Labels the microtubule network. 1:500 dilution, fixation with methanol or paraformaldehyde (PFA) 3%.
Anti-Cep55	Rabbit	Mouse	ab170414, Abcam	Labels the microtubule network. 1:500 dilution, fixation with PFA 3%.
Anti-γ-tubulin	Rabbit	Mouse	ab179503, Abcam	Labels the microtubule organizing center (MTOC) or centrosome. 1:500 dilution, fixation with methanol.

Table 1 Antibodies and reagents.—cont'd

Primary/ secondary antibody	Host	Reactivity	Catalog ID	Observations
Anti-pericentrin	Rabbit	Mouse	ab4448, Abcam	Labels the microtubule organizing center (MTOC) or centrosome. 1:500 dilution, fixation with methanol.
Anti-LAMP1	Rat	Mouse	553,792, BD Pharmigen	Labels lysosomes. 1:200 dilution, fixation with PFA 3%.
AffiniPure F(ab')$_2$ Fragment Alexa Fluor 488, IgG (H+L)	Donkey	Rabbit	711-546-152, Jackson Immunoresearch	1:500 dilution.
AffiniPure F(ab')$_2$ Fragment Cy3, IgG (H+L)	Donkey	Rabbit	711-166-152, Jackson Immunoresearch	1:500 dilution.
AffiniPure F(ab')$_2$ Fragment Alexa Fluor 647, IgG (H+L)	Donkey	Rabbit	711-606-152, Jackson Immunoresearch	1:500 dilution.
AffiniPure F(ab')$_2$ Fragment Alexa Fluor 488, IgG (H+L)	Donkey	Rat	712-546-150, Jackson Immunoresearch	1:500 dilution.
AffiniPure F(ab')$_2$ Fragment Cy3, IgG (H+L)	Donkey	Rat	712-166-153, Jackson Immunoresearch	1:500 dilution.
AffiniPure F(ab')$_2$ Fragment Alexa Fluor 647, IgG (H+L)	Donkey	Rat	712-606-153, Jackson Immunoresearch	1:500 dilution.
Hoechst			H3570, Invitrogen	Labels the nucleus. 1:1000 dilution, fixation with PFA.

Continued

Table 1 Antibodies and reagents.—cont'd

Primary/ secondary antibody	Host	Reactivity	Catalog ID	Observations
Rhodamine phalloidin			R415, Invitrogen	Labels polymerized actin. 1:400 dilution, fixation with PFA.
NH-2 Beads			Polysciences, Inc. Cat. No. 17145-5	Consider the size of the beads depending on the cell that will interact with them. In our model, we use $3\,\mu m$ size for beads.
Fluoromount G			Electron Microscopy Sciences, Cat No. 17984-25	Add $5\,\mu L$ for $10\,mm$ coverslips and $7\,\mu L$ for $12\,mm$ coverslips.
Microscope Cover Glasses			Paul Marlenfeld Gmb H & Co, Ref 0111500	If super-resolution techniques will be used, consider alternatives more suitable with the optical configuration.

1.1. First, it is recommended to crop individual cells using the cell_cutter. ijm (drag and drop into Fiji main window) macros available on the GitHub repository. Instructions are available in the code.

1.2. Open the image and define the bead and cell areas to analyze using the circle tool selection to delimit both boundaries and save them as regions of interest (ROI) (Fig. 1). If the cell has an irregular shape, the freehand tool can be used to select the area. For cell outline delimitation, both actin staining or bright field can be used.

1.3. Determine the cell center (**CC**) and the bead center (**BC**) by running **Analyze | Measure** on cell and bead areas, respectively. The **X and Y** values obtained from the **Results** window determine the center coordinates, as highlighted in the figure with a red dashed box (Fig. 1).

Quantitative imaging of the B cell synapse 239

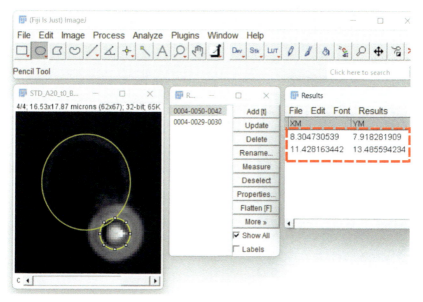

Fig. 1 Screenshot of ImageJ showing Results (highlighted in the figure with a red dashed box) of center coordinates of both the cell and bead ROIs, after running Analyze | Measure.

1.4. Manually determine the center of the **centrosome** (highest intensity of the antibody staining) using the **Point tool** selection in ImageJ and run **Analyze | Measure**. The **X** and **Y** values obtained from the **Results** window determine the centrosome coordinates, as highlighted with a red dashed box in the figure (Fig. 2).

1.5. Then, draw an angle from **CC** to **Cell structure** (**a**) and from **CC** to **BC** (**b**) using the **Angle tool** selection and run **Analyze | Measure**. The **Angle** value (highlighted with a red dashed box in the figure) in the **Results** window shows the angle (α) between both vectors (**a** and **b**) (Fig. 3).

1.6. Calculate the **Polarity Index** using the following formula:

$$\text{Polarity Index} = a \frac{\cos(\alpha)}{b}$$

1.7. Representative results

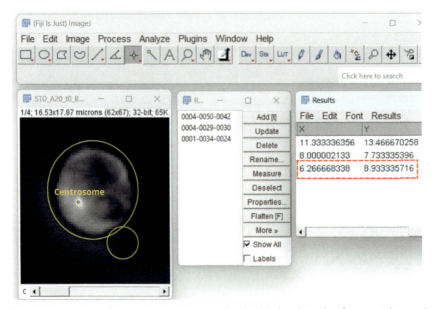

Fig. 2 Screenshot of ImageJ showing Results (highlighted in the figure with a red dashed box) of coordinates the cell centrosome, after running Analyze | Measure.

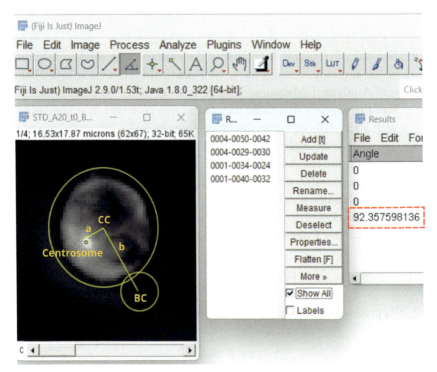

Fig. 3 Screenshot of ImageJ showing the Result (highlighted in the figure with a red dashed box) of the value of the angle (α) formed by the vectors a (cell center—centrosome) and b (cell center—bead center), after running Analyze | Measure.

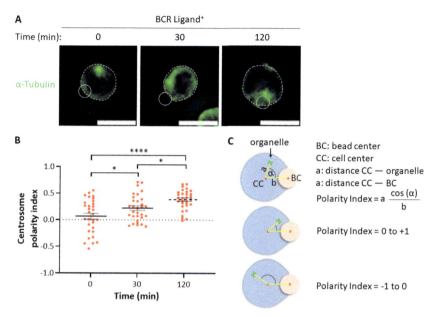

Fig. 4 (A) Images of B cells stained for α-tubulin (shown in green) activated with BCR ligand–positive beads for the indicated times. Scale bar: 10 μm. (B) Centrosome polarity index calculated from the images in (A). *$P<0.05$; ****$P<0.0001$; one-way ANOVA with Holm-Sidak's multiple comparison test; $n \geq 31$ cells from three independent experiments. (C) Scheme depicting how to calculate the polarity index of organelles (centrosome or Golgi apparatus) toward the IS. a = distance between the center of the cell (CC) to the organelle. b = distance from the CC and to the bead center (BC).

The following image shows the centrosome recruitment to the IS upon B cell stimulation with antigen-coated beads over time (Fig. 4).

2. Estimating the polarity index for lysosomes

We use this algorithm to analyze the polarity of organelles that display a more dispersed distribution. In this case we have labeled lysosomes with Lamp1 (see Table 1).

2.1. Define the bead and cell areas to analyze, using the **Circle tool** selection to delimit both boundaries and save them as ROI (Fig. 5). Once the bead and cell areas have been determined, set the fluorescence channel and project the image into one z-stack (**Image | Stacks | Z-Project [Sum slices]**), then run **Analyze | Measure** and extract the mass center (**MC**) coordinates (**XM** and **YM**) from the Results windows, these values are highlighted with a red dashed box in the figure (Fig. 6).

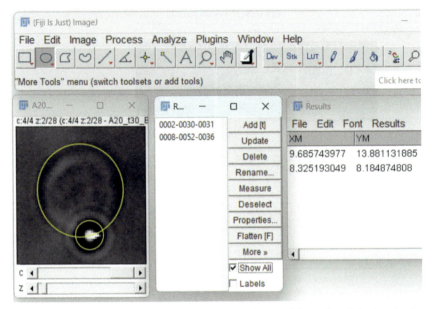

Fig. 5 Screenshot of ImageJ showing the delimited area of the cell and the bead using the circle tool in the bright field channel.

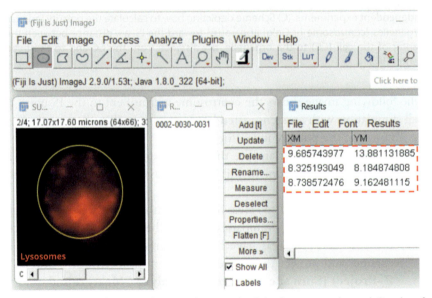

Fig. 6 Screenshot of ImageJ showing the z-stack of the lysosomes channel. Results of the cell center, bead center and mass center coordinates (highlighted with a red dashed box) after running Analyze | Measure, are shown.

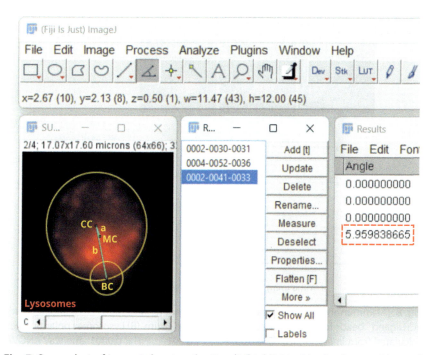

Fig. 7 Screenshot of ImageJ showing the Result (highlighted in the figure with a red dashed box) of the value of the angle (α) formed by the vectors a (CC—MC) and b (CC—BC), after running Analyze | Measure.

2.2. Apply the same algorithm mentioned before changing **Cell structure** for **MC**. Thus, the angle (**α**) is defined by CC-MC (**a**) and CC-BC (**b**). The angle value is highlighted with a red dashed box in the figure (Fig. 7).

2.3. Calculate the polarity using the following formula:

$$\text{Polarity Index} = a \frac{\cos(\alpha)}{b}$$

2.4. Representative results

The following image depicts the quantification of polarized organelles that display a dispersed distribution, such as lysosomes (Fig. 8). In this case, we use the lysosome center of mass coordinates (Fig. 6). The images show that the polarity indexes of lysosome pools reach more positive values over time, indicating that lysosomes are being recruited to the IS during B cell activation (Fig. 8).

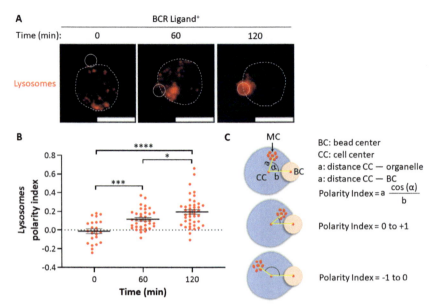

Fig. 8 (A) Immunofluorescence staining of Lysosomes (LAMP1) in B cells incubated with BCR ligand[+] beads for the indicated times. Scale bar: 10 μm. (B) Lysosomes polarity index calculated from images in (A) $*P < 0.05$; $***P < 0.001$ $****P < 0.0001$; one-way ANOVA with Holm-Sidak's multiple comparison test; $n \geq 23$ cells from three independent experiments. (C) Scheme depicting how to calculate the polarity index of lysosomes toward the IS. a = distance between the center of the cell (CC) to the mass center (MC). b = distance from the CC and to the bead center (BC).

5. Characterization of the immune synapse organization of B cells activated on antigen-coated surfaces

This analysis is used to quantify organelles that are recruited to the IS of B cells activated on an antigen-coated surface. In this example, we will quantify the lysosomes at the IS as a marker of B cell activation. This analysis could also be used to segment and quantify other discrete markers, such as Actin foci at the IS.

1. Measuring organelle recruitment at the IS area
    1.1. As a requisite, you must previously cut single cells from the original images. An extra macro is provided for this task (cell_cutter.ijm in GitHub repository). **See Note 12**.
    1.2. The macros will prompt you to select the channel of interest, in this case, we will focus on lysosomes.

1.3. Then, determine the contact slice between the cell and the surface (IS) and define the cell outline with a cytoskeleton marker (actin channel in the example). You could use the freehand or circle tool selection to delimit the boundaries (Fig. 9).

1.4. Take into consideration the Z plane of interest. The macros will ask you for this number. In this case, it is number 6 (z:6/35 in the image title). Only use entire numbers in this section.

1.5. Next, the macros will convert the image to 8-bit and the previously outlined cell ROI will be analyzed with "Analyze Particles." We use "size = 0.10–4.00 μm^2" for lysosome analysis (de Araujo, Liebscher, Hess, & Huber, 2020). In Fig. 10, LAMP1 signal and segmented particles are shown (left and right images, respectively). **See Note 13**.

1.6. The next image in the folder will be opened automatically to proceed with the same steps as before.

1.7. Finally, Fiji's—ImageJ "Log" shows the number and name of images, the area of the cell, the number and area of particles, and the average size of them within the cell. Save this table with results.

1.8. Representative results

The following image depicts the quantification of particles present at the IS. In this case, we measured the number of lysosomes recruited at the IS plane (Fig. 11D) and the formation of actin foci at the IS (Fig. 11B). The images show that the number of lysosome and actin foci at the IS increase over time, indicating the state of activation of the B cell (Fig. 11D and C, respectively).

2. Measurement of organelle distribution at the IS

After the activation of B cells, lysosomes are preferentially found in the center of the IS. This feature is often referred to as a marker of activation

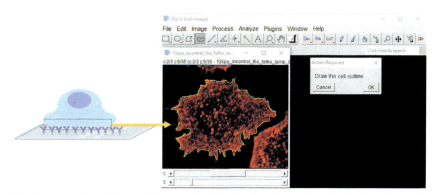

Fig. 9 Screenshot of ImageJ showing the freehand tool to select the border of the cell.

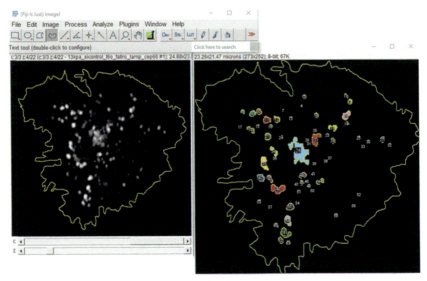

Fig. 10 Screenshot of ImageJ showing the LAMP1 signal and segmented particles are shown.

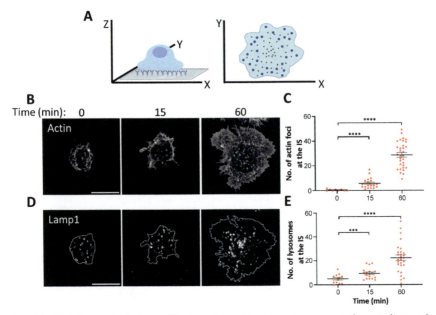

Fig. 11 (A) Scheme depicting activation of B cells onto antigen-coated coverslips and representative image of YX planes. (B–D) Representative confocal images of actin (Phalloidin) and Lamp1 (lysosomes) in resting (0 min) and activating (15–60 min) conditions. (C) Quantification of the number of actin foci at the IS at different conditions. (E) Quantification of the number of lysosomes at the IS. $N > 10$ cells. $P < 0.0001$; one-way ANOVA with Tukey's multiple comparisons tests. Scale bar = 10 um.

in B cells, along with other organelles and structures, such as the BCR. To analyze organelle distribution at the IS, we make available the following ImageJ macros (2D_distribution_analysis.ijm in the GitHub repository). The following instructions serve as a guide for using and understanding its components. Importantly, the results of these macros can be further processed with an R-language script that will provide the user with a final .csv file containing the values of the central accumulation of a marker of interest (Rscript_2D_analysis.R in the GitHub repository).

2.1. Previously cut single cells from the original images. An extra macros is provided for this task (cell_cutter.ijm in the GitHub repository). In addition, generate two folders to organize the images and the results. We recommend naming them "cells" and "results."

2.2. Run the macros 2D_distribution_analysis.ijm. Before starting the analysis, determine the contact slice and define the cell outline, and press "Ctrl + t."

2.3. Automatically, the macros will modify the scale of the ROI to the factor you choose. In this example, the ROI was rescaled to 0.4 times the size of the total outline of the cell (center). The ROI will be analyzed with "Analyze Particles." **See Note 14**.

```
50   roiManager("Select", 1);
51   RoiManager.scale(0.4, 0.4, true);
52   roiManager("Select", 0);
53   roiManager("Select", newArray(0,1));
54   roiManager("XOR");
55   roiManager("Add");
56   roiManager("Select", 2);
57   roiManager("Delete");
58   roiManager("Select", newArray(0,1,2));
59   roiManager("multi measure");
```

2.4. Excel files will be generated for each cell analyzed and automatically saved in the results folder. The respective ROIs of each cell will also be saved.

2.5. Using the R-script available at the repository, you can generate a final Excel-compatible file that reports the center/periphery ratio of accumulation for each channel present on the images. Calculate the **Ratio** using the following formula:

$$Ratio\ (r) = \frac{MeanFluorescentIntensity\ (MFI)\ center}{MFI\ periphery}$$

2.6. Representative results

The following image (Fig. 12) depicts the quantification of particles distributed at the IS. In this case, we analyzed lysosome distribution in the IS plane. The images show that lysosomes are equally distributed in the IS plane at early times of activation, but after 60 min of B cell activation they are accumulated at the center at the IS.

3. Measuring organelle distribution in Z planes

This analysis determines the general distribution of fluorescence intensity across the Z-slide of cells seeded over coverslips. We use this measurement to assess the organelle proximity to the immune synapse.

3.1. To quantify the fluorescence distribution in the Z-slides, first determine the contact slide and then the plane corresponding to the upper limit of the cell.

Note: It is recommended to cut single-cell images from the contact plane to the top edge of the cell. You can analyze different cells in the same image if all cells are found in the same contact (lower) and upper planes.

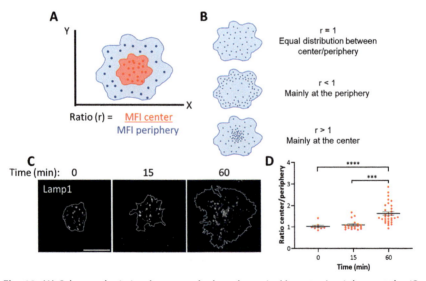

Fig. 12 (A) Scheme depicting how to calculate the ratio (r) center/periphery at the IS. (B) Schematic representation of mean fluorescent intensity (MFI) distribution and r interpretation. (C) Representative confocal images of Lamp1 in resting (0 min) and activating (15–60 min) conditions. (D) The respective quantification of lysosome accumulation at the center and periphery. $N > 10$ cells. $P < 0.0001$; one-way ANOVA with Tukey's multiple comparisons tests. Scale bar = 10 μm.

3.2. Draw the cell outline at the contact plane and add the area to the ROI manager with "Ctrl + t."
3.3. At the ROI manager, click on "Select All" areas and analyze them with the "multi-measure" option.
3.4. Save the results. From the results table you can filter all measurements, in this case, the Mean intensity of fluorescence and Z-slides of the cell. You can select a channel of interest and filter the data.
3.5. Representative results

The following image depicts the quantification of centrosome polarization in B cells activated on antigen-coated surfaces (Fig. 13A). By measuring the fluorescence intensity of the centrosome in all z-slices of the cell we observed that its highest peak of intensity is closer to the IS in activated B cells, while it is equally distributed among z-slices at early times of activation (Fig. 13B and C).

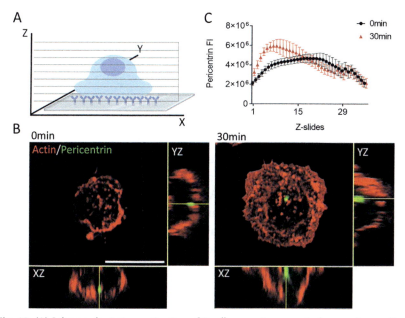

Fig. 13 (A) Scheme depicting activation of B cells on antigen-coated coverslips and how to determine the distribution of the signal of interest in the Z plane, and the plot per Z fraction. (B) Representative confocal images of B cells labeled actin and pericentrin (centrosome) activated onto Ag-coated coverslips in different conditions, showing the YZ and XZ planes. (C) The distribution of pericentrin across the Z plane is represented by a line plot of the fluorescence intensity distribution versus each Z fraction. $N = 5$ cells. Mean with SEM are shown.

6. Discussion and relevance

In conclusion, the analyses presented here represent a significant advancement in analyzing the immune synapse of B cells. By providing detailed information on the spatiotemporal organization of this crucial structure, these methods allow for a better understanding of the mechanisms underlying B cell activation and provides a mean to analyze the effects of extracellular cues on B cell function (Lagos et al., 2022). It is important to note, however, that the quality of microscopy imaging is critical to the success of these methods. Any deviation in sample preparation or imaging settings could significantly impact the results.

These methods are compatible with open-source applications, providing accessibility to researchers worldwide. In addition, the raw data derived from the results can be explored to perform further analyses that are not considered here, for example, the study of antigen extraction, processing, and presentation by quantitative measurements of protein recruitment and tracking of intracellular vesicles. Therefore, these methods have the potential to not only contribute to a deeper understanding of the immune synapse of B cells but also in other cell models.

7. Notes

1. The centrosome of B cells can be identified by transfecting cells with a centrin–GFP expression plasmid and actin cytoskeleton can be detected by transfecting cells with a LifeAct expression plasmid. Depending on the needs of the experimental setup, one can use specific antibodies such as gamma tubulin for a discrete labeling of the centrosome. If not available, one can use alpha tubulin and consider the highest intensity point as the MTOC (centrosome).
2. To avoid the binding of ligands to the Fc receptor, use F(ab′) or F(ab′)$_2$ antibody fragments instead of full-length antibodies.
3. Antigen-coated beads should not be stored for more than 1 month.
4. Consider preparing the coverslips the day before the assay. UV lamps or plasma can be used to clean the coverslip before use. To assess the ligand activating density on a surface one could label the ligand with a fluorochrome and calculate the mean fluorescence intensity on a region (or bead, depending on the experiment) to ensure homogenous coating.

5. It is important to mix by vortex instead of pipetting up and down because this could reduce the number of beads in the sample, due to their accumulation in the plastic tip of the pipette. Use beads coated with BCR-ligand⁻ as a negative control for each assay. We recommend starting the activation with the longest incubation time and proceeding with the following time points. For shorter times of activation (0 min) you can add thrice the cells, and thrice the amount of beads, meaning 450,000 of cells with 1350,00 of beads to have a sufficient amount of adhering cells on the coverslip. Calculate intervals of activation for samples such that they are ready for fixation at the same time.
6. To decide which fixation method to use, check the antibody/dye data sheet.
7. At this point, coverslips can be stored in 1× PBS in a humid chamber sealed at 4 °C for a maximum of 3 days. In the image below we illustrate two options for using and crafting the humid chamber. In (a) a drop of PBS or antibody solution lays on top of the coverslip, or, as shown in (b) the coverslip is placed on top of the drop.

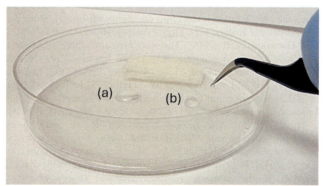

8. To avoid cross-reactivity of the secondary antibody with the B cell receptor of B cells, do not use mouse-derived primary antibodies unless they are fluorescence labeled primary antibodies.
9. Consider using an "anti-fade" mounting reagent (see the Table of Materials).
10. When using a confocal microscope, we recommend taking 0.2 μm thick stacks when acquiring three-dimensional (3D) images. This allows the analysis of the cell using z-stacks analysis
11. This algorithm can be used for cell structures or organelles observed as discrete markers.

12. It is recommended to analyze individual cells because the plane of interest may not be the same for all cells. In the GitHub repository the user can find examples for testing the macros and scripts provided.
13. You can modify the particle size depending on your interests. For example, we use "size $= 0.10$–$0.50\,\mu m^2$" for actin foci analysis (Roper et al., 2019).
14. You can modify the ROI scaling depending on your interests. In this case, we use the 0.4 scale factor to analyze a specific central region and to discriminate organelles separated from actin enriched areas. The previous is based in the knowledge that the immune synapse of B cells is characterized by three concentric regions: the central supramolecular activation cluster (cSMAC) in which BCRs–ligands are concentrated, the peripheral SMAC (pSMAC), which contains adhesion molecules such as LFA1, and the distal SMAC (dSMAC) in which actin is enriched.

References

Arena, E. T., Rueden, C. T., Hiner, M. C., Wang, S., Yuan, M., & Eliceiri, K. W. (2017). Quantitating the cell: Turning images into numbers with ImageJ. *Wiley Interdisciplinary Reviews: Developmental Biology, 6*(2). https://doi.org/10.1002/wdev.260.

Batista, F. D., & Harwood, N. E. (2009). The who, how and where of antigen presentation to B cells. *Nature Reviews. Immunology, 9*(1), 15–27. https://doi.org/10.1038/nri2454.

de Araujo, M. E. G., Liebscher, G., Hess, M. W., & Huber, L. A. (2020). Lysosomal size matters. *Traffic, 21*(1), 60–75. https://doi.org/10.1111/tra.12714. Blackwell Munksgaard.

Lagos, J., et al. (2022). Autophagy induced by toll-like receptor ligands regulates antigen extraction and presentation by B cells. *Cell, 11*(23). https://doi.org/10.3390/cells11233883.

Lankar, D., et al. (1998). *Syk tyrosine kinase and B cell antigen receptor (BCR) immunoglobulin-subunit determine BCR-mediated major histocompatibility complex class ii-restricted antigen presentation.* [Online]. Available: http://www.jem.org.

Reversat, A., et al. (2015). Polarity protein Par3 controls B-cell receptor dynamics and antigen extraction at the immune synapse. *Molecular Biology of the Cell, 26*(7), 1273–1285. https://doi.org/10.1091/mbc.E14-09-1373.

Roper, S. I., Wasim, L., Malinova, D., Way, M., Cox, S., & Tolar, P. (2019). B cells extract antigens at Arp2/3-generated actin foci interspersed with linear filaments. *eLife, 8*, e48093. https://doi.org/10.7554/eLife.48093.

Yuseff, M.-I., Pierobon, P., Reversat, A., & Lennon-Duménil, A.-M. (2013). How B cells capture, process and present antigens: A crucial role for cell polarity. *Nature Reviews. Immunology, 13*(7), 475–486. https://doi.org/10.1038/nri3469.

Yuseff, M. I., et al. (2011). Polarized secretion of lysosomes at the B cell synapse couples antigen extraction to processing and presentation. *Immunity, 35*(3), 361–374. https://doi.org/10.1016/j.immuni.2011.07.008.

CHAPTER TWELVE

γδ T cell expansion and their use in *in vitro* cytotoxicity assays

María Alejandra Parigiani[a,b,c,*]

[a]Faculty of Biology, University of Freiburg, Freiburg, Germany
[b]Signalling Research Centres BIOSS and CIBSS, University of Freiburg, Freiburg, Germany
[c]Center of Chronic Immunodeficiency (CCI) and Institute for Immunodeficiency, University Clinics and Medical Faculty, Freiburg, Germany
*Corresponding author. e-mail address: maria.parigiani@bioss.uni-freiburg.de

Contents

1. Introduction	254
2. Graphical abstract	256
3. Materials	256
3.1 Equipment	256
3.2 Disposable materials	256
3.3 Chemicals and biological products	257
4. Methods	258
4.1 Culturing of patient-derived BCSC	258
4.2 Transfection of BCSC	258
4.3 Primary γδ T cell isolation and expansion and culture conditions	258
4.4 MAC sorting and determination of purity of γδ T cells by flow cytometry	259
4.5 Coculture with target cells and cytotoxicity assessment via luminescence	259
5. Concluding remarks	262
6. Notes	262
Acknowledgments	263
Declaration of interests	263
References	263

Abstract

Limited therapeutic options for triple-negative breast cancer (TNBC) patients prompted the exploration of advanced immunotherapeutic approaches in this cancer entity. γδ T cells started gaining attention for their remarkable ability to suppress skin cancer, which rapidly extended to other cancer entities. This special T cells represent a suitable immune population to be used in adoptive T cell transfer approaches. Combining characteristics of both αβ T cells and natural killer (NK) cells, these unique T cells exhibit swift cancer cell elimination independent of MHC class I antigen presentation. The distinct advantage of γδ T cell immunotherapy lies in its HLA-unrestricted nature, enabling the utilization of cells from healthy donors. Up to date, many studies demonstrate that also expanded γδ T cells from breast cancer patients exhibit enhanced cytotoxicity and cytokine release in vitro,

Methods in Cell Biology, Volume 193
ISSN 0091-679X
https://doi.org/10.1016/bs.mcb.2024.05.002

Copyright © 2025 Elsevier Inc.
All rights are reserved, including those
for text and data mining, AI training,
and similar technologies.

paving the way for γδ T cell-based therapies. The approach outlined below offers an alternative method for conducting in vitro cytotoxicity assays, utilizing γδ T cells as the effector cell population and breast cancer stem cells as the target.

Abbreviations

ACT	adoptive T cell transfer
BCSC	breast cancer stem cells
ER	estrogen receptor
E:T	effector to target ratio
HD	healthy donor
HER-2	human epidermal growth factor receptor
IL2, IL4	interleukin 2, interleukin 4
MACS	magnetic-activated cell sorting
PBMCs	peripheral blood mononuclear cells
PR	progesterone receptor
TCR	T cell-receptor
TNBC	triple-negative breast cancer

1. Introduction

Breast cancer (BC) is frequently acknowledged as the most common cancer type worldwide and is certainly the most common cancer type among women, with the highest mortality rate. It manifests as a heterogeneous disease categorized into molecular subtypes based on distinct molecular features, providing a basis for tailored treatment strategies. These subtypes, determined by immunohistochemical expression of hormone and other receptors, include Luminal A (estrogen receptor (ER) positive and/or progesterone receptor (PR) positive), Luminal B (ER positive, mostly PR negative, Ki67 high), human epidermal growth factor receptor (HER2) positive, and triple-negative breast cancer (TNBC), which may also harbor mutations such as BRCA1 or BRCA2 (Smolarz, Nowak, & Romanowicz, 2022).

Adoptive T cell transfer (ACT), a highly promising anticancer therapy, harnesses the cytotoxic activity of a patient's own T cells to combat cancer. In addition to αβ T cells, γδ T cells, positioned at the interface of innate and adaptive immunity, have garnered attention as a viable immune population for immunotherapies since their demonstrated efficacy in suppressing cutaneous malignancies in 2001 (Girardi et al., 2001). γδ T cells have demonstrated effective targeting of cancer cells and a positive correlation between infiltration and prognosis in a plethora of cancer entities, including cancers of the gastro intestinal tract, hepatocellular carcinoma, triple-negative breast cancer, ovarian cancer, renal cell carcinoma, and glioblastoma, among others (Andreu-Ballester et al., 2020; Boissière-Michot et al., 2021; Bryant et al., 2009; Foord, Arruda, Gaballa, Klynning, & Uhlin, 2021; Wang et al., 2017; Zhao et al., 2021). This has led to their use in immunotherapies and in ongoing clinical trials as well as earning a FDA fast track designation for the use of γδ-CAR T cells in the treatment of relapsed or refractory B-cell non-Hodgkin lymphoma in 2022 (Nishimoto et al., 2022; Saura-Esteller et al., 2022).

In humans, γδ T cells are categorized into three major subsets based on their TRDV genes, known as Vδ1+, Vδ2+, and Vδ3+. Vδ1+ cells predominantly reside in the gut epithelium, skin, spleen, and liver, with only a minimal presence in the circulating blood (Hu et al., 2023). Conversely, the Vδ2+ subset predominantly pairs with the Vγ9 TCR, making it the dominant γδ T cell population in the bloodstream. While Vδ1+ T cells demonstrate adaptive immune characteristics such as "memory-like" features and rapid clonal expansion capacity, the semi-invariant Vγ9Vδ2 T cells align more closely with innate immunity, since this population gets activated through the recognition of phosphoantigens, mediated through BTN3A isoforms (Adams, Gu, & Luoma, 2015). The Vδ3+ subset is seldomly detected in the peripheral blood of healthy individuals, being almost exclusive to liver and gut epithelium. Interestingly, Vδ3+ cells exhibit recognition of similar ligands as the Vδ1+ subset, as both have been found to recognize CD1-d proteins, which play a role in the presentation of lipid, lipopeptide, and glycolipid antigens (Deseke & Prinz, 2020). While there are many known ligands for Vδ1+, most Vδ3+ ligands remain undiscovered. The unique antitumor mechanisms of γδ T cells, integrating features of both αβ T cells and natural killer (NK) cells, rely not only on the T cell receptor (TCR) and co-stimulatory signals but also on NK cell receptors. This dual mechanism enables rapid elimination of aberrant cells while preserving normal tissue, independent of MHC class I antigen presentation (Mensurado, Blanco-Domínguez, & Silva-Santos, 2023). The distinctive characteristic of γδ T cell immunotherapy lies in its HLA-unrestricted nature, setting it apart from other cell-based therapeutic approaches. This feature enables the utilization of cells sourced from healthy donors, which can undergo in vitro expansion through various protocols. Subsequently, these expanded cells can be transferred to patients, akin to the process employed in other T cell transfer methodologies (Siegers, Ribot, Keating, & Foster, 2013).

The approach detailed here, involving the efficient in vitro eradication of breast cancer stem cells (BCSC) by ex vivo expanded γδ T cells, was documented by Raute et al. (2023). This protocol employs the expansion procedure established by Siegers et al., enabling the amplification of the Vδ1+ and Vδ2+ subsets of γδ T cells (Siegers et al., 2013).

2. Graphical abstract

See Fig. 1.

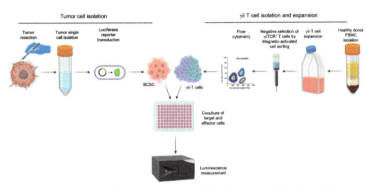

Fig. 1 Graphical abstract of the expansion of γδ T cells and their use in cytotoxicity assays against BCSC in vitro. *Created with BioRender.com.*

3. Materials

3.1 Equipment

1. Hypoxia incubator chamber (Stem cell technologies, Cat no. 27310)
2. Flow cytometer (Invitrogen™, Attune™)
3. Multimode microplate reader capable of recording luminescence (Molecular devices, SpectraMax)
4. Biological safety cabinet (Thermo Scientific™ Herasafe™) and cell culture incubators (Thermo Scientific™ Heracell™ CO_2 Incubators)

3.2 Disposable materials

1. 6 well plates for suspension cell culture, transparent (Sarstedt)
2. Sterile microcentrifuge tubes (Eppendorf)
3. 96 well plates for cell culture round and flat bottom, transparent (Greiner)
4. 96 well plates flat bottom, white (Perkin Elmer)
5. 10 cm cell culture plates for adherent cell culture

3.3 Chemicals and biological products

1. Mammary stem cell (MSC) medium: MEBM medium (Lonza, CC-3151) supplemented with $1\times$ B27 (Thermo Fisher Scientific, 17504-044), $1\times$ amphotericin (Sigma-Aldrich, A2942), 20 ng/mL EGF (Peprotech, AF-100-15B), 20 ng/mL FGF (Peprotech, AF-100-18B), 4 µg/mL heparin (Sigma-Aldrich, H3149), 35 µg/mL gentamicin (Thermo Fisher Scientific, 15750-045), 500 nmol/L H-1152 (Calbiochem, 555552), 100 µg/mL penicillin, and 100 µg/mL streptomycin (Thermo Fisher Scientific, 15140-122).
2. InSolution Rho Kinase Inhibitor (Merck Millipore, H-1152)
3. Matrigel, Growth Factor Reduced (Corning; #354230)
4. Accutase (Sigma-Aldrich, #A6964-100ML)
5. Buffy coats were obtained from banked whole blood of healthy donors
6. Pancoll human (Pan Biotech, P04-601000)
7. $\gamma\delta$ T-cell medium: RPMI1640 medium (Life Technologies, 11554516) supplemented with 10% FBS, 100 µg/mL penicillin, 100 µg/mL streptomycin, 10 mmol/L HEPES, 10 µmol/L sodium pyruvate, and $1\times$ MEM nonessential amino acids (Pan Biotech, P08-32100).
8. Recombinant human IL-2 IL2 (Peprotech, 200-02)
9. Recombinant human IL4 (Peprotech, 200-04)
10. Dulbecco's Phosphate Buffered Saline (Anprotec, #AC-BS-0002)
11. Concanavalin A (Sigma-Aldrich, C5275)
12. pCMV ΔR8.74 and pMD2.vsvG, lentiviral packaging plasmid and lentiviral envelope for virus particle generation
13. Luciferase reporter plasmid for lentiviral transduction (e.g., pHRSIN-CS-Luc-IRES-emGFP)
14. D-Luciferin Firefly (Biosynth, #FL08608)
15. MACS TCRγ/δ + T Cell Isolation Kit, Miltenyi Biotec, 130-092-892 and LS Columns for MAC Sorting of $\gamma\delta$ + T cells
16. In case of Vδ1 and Vδ2 isolation as performed for Fig. 3, Biotin-TCR Vδ2 Antibody (anti-human, Biolegend, #331404 or Biotin-TCR Vδ1 Antibody (anti-human, REAfinity™, Miltenyi Biotec, #130-120-439) were used, depending on the desired population in negative selection
17. Antibodies to check $\gamma\delta$ + T culture purity: PE-$\gamma\delta$ TCR (anti-human, Life Technologies, #MHGD04), APC-TCR Vδ1 Antibody (anti-human, REAfinity™, Miltenyi Biotec, #130-118-968), BV421-TCR Vδ2 Antibody (anti-human, BioLegend, #331428)
18. Gas composition: 92% N_2, 5% CO_2, 3% O_2 for BCSC

19. Zoledronate (Zoledronic acid monohydrate) (Sigma-Aldrich, SML0223-10MG)
20. Mevastatin (M2537-5MG)

4. Methods
4.1 Culturing of patient-derived BCSC
BCSC were generated from independent breast tumor samples, resembling the TNBC subtype.
1. Tumors were resected and cells mechanically dissociated as previously described (Metzger et al., 2017)
2. Upon dissociation, cells were mixed 1:1 with Matrigel and filled with complete MSC medium adding fresh 500 nM Rho kinase inhibitor
3. BCSCs were cultured at 37 °C under low oxygen conditions using a hypoxia chamber and a gas mix (3% O_2, 5% CO_2, 92% N_2)
4. Cells were passaged every 2 days and passaging was performed as previously described (Metzger et al., 2017)

4.2 Transfection of BCSC
The pHRSIN-CS-Luc-IRES-emGFP luciferase reporter plasmid was previously generated by Gibson assembly as described by Dang et al. (2021).
1. Lentiviral particles were produced by plating 1×10^7 HEK293 cells and transfecting them with the pHRSIN-CS-Luc-IRES-emGFP luciferase reporter plasmid using the pCMV ΔR8.74 and pMD2.vsvG lentiviral packaging- and envelope plasmid for virus particle generation
2. BCSC cells were plated and transduced the subsequent day with concentrated HEK293 supernatants
3. GFP positive BCSC were isolated via fluorescence cell sorting and cultured as described above

4.3 Primary γδ T cell isolation and expansion and culture conditions
1. Peripheral blood mononuclear cells (PBMCs) from healthy donors (HD) banked blood were isolated by conventional density gradient centrifugation (Pancoll human, Pan Biotech, P04-601000) as described previously (Siegers et al., 2013)
2. Upon isolation, PBMCs were resuspended at a concentration of 1×10^6 cells/mL in γδ T-cell medium

3. γδ T-cell expansion was induced by adding Concanavalin A to a final concentration of $1\,\mu g/mL$, together with IL2 and IL4, to a final concentration of $10\,ng/mL$ each
4. Cell density was adjusted every 48h to 1×10^6 cells/mL by adding γδ T-cell medium, supplemented with fresh IL2 and IL4 throughout the time they were kept in culture
5. γδ T-cells were cultured at $37\,°C$ in a humidified atmosphere of 5% CO_2

4.4 MAC sorting and determination of purity of γδ T cells by flow cytometry

1. Around day 14, γδ T cells were isolated from the PBMC culture using negative selection following the manufacturers protocol (MACS TCRγ/δ + T Cell Isolation Kit, Miltenyi Biotec, 130-092-892)
2. Purity yield of total γδ T cells, and Vδ1 and Vδ2 populations present in the culture were determined by flow cytometry using PE-γδ TCR, APC-TCR Vδ1 Antibody, BV421-TCR Vδ2 Antibody in a 1:200 dilution
3. Cultures with a purity of $\geq 90\%$ were used for experiments from day 14 on until day 21 post expansion begin
4. Cell density was adjusted every 48h to 1×10^6 cells/mL by adding γδ T-cell medium, supplemented with fresh IL2 and IL4 throughout the time they were kept in culture, and experiments were performed in between the cytokine supplementation for consistency
5. For further isolation of Vδ1 or Vδ2 γδ T-cell populations, we used MACS negative selection and added a biotinylated antibody contrary to the population we aimed to isolate to the first step/primary antibody mix of the MACS isolation (Biotin-TCR Vδ1 Antibody was used when aiming to isolate Vδ2 and Biotin-TCR Vδ2 Antibody was used when aiming to isolate Vδ1 γδ T-cell), using $10\,\mu L$ antibody per 10^7 cells (Fig. 2)

4.5 Coculture with target cells and cytotoxicity assessment via luminescence

Cytotoxicity assessment was performed as previously described (Hartl et al., 2020).

1. 1×10^4 luciferase-expressing BCSCs were plated in triplicates in a white 96-well flat-bottom plate
2. $37.5\,\mu g/mL$ D–Luciferin Firefly was previously added to the cell suspension

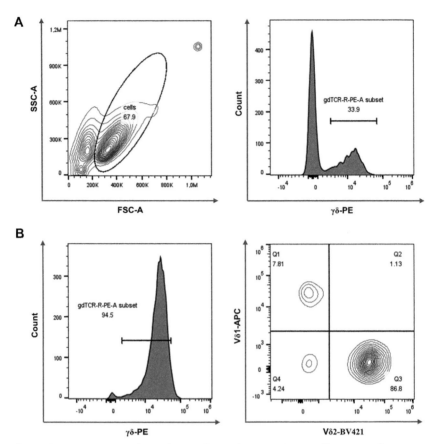

Fig. 2 Flow cytometry analysis of γδ T cell populations. (A) γδ T cell staining after expansion, prior to MACS. (B) γδ T cell populations Vδ1 and Vδ2 after γδ T cell enrichment by MAC sorting.

3. Effector cells were added to the target cells at the desired effector to target cell ratios (e.g. 10:1 E:T)
4. As spontaneous cell death controls, target cells without effector cells were used. Maximum killing controls were made using target cells lysed with 1% Triton X-100
5. Bioluminescence was measured as relative light units (RLU) and measurements were carried out using a multimode microplate reader (SpectraMax) at several time points (0, 4, 8, 12, 16, 18, 24 h), starting at 0 h after adding effector cells to target cells
6. The percentage of specific lysis was calculated as follows (Fig. 3):

$$\% \text{specific lysis} = 100 \times \frac{\text{spontaneous death RLU} - \text{test RLU}}{\text{spontaneous death RLU} - \text{Maximum killing RLU}}$$

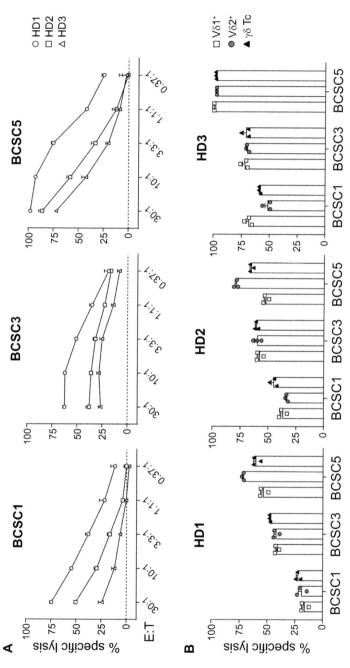

Fig. 3 In vitro expanded γδ T cells kill BCSC. (A) Target cells from different breast cancer donors transduced with Luciferase were killed by in vitro expanded γδ T cells from three healthy donors. The figure depicts different effector to target ratios after 8h of initiating the coculture (means±SEM). (B) Specific lysis (means±SEM) achieved using total Concanavalin A-expanded γδ T cells (γδ Tc) and MACS-isolated Vδ1 and Vδ2 and after MACS from different healthy donors at an E:T ratio of 10:1 (Raute et al., 2023).

5. Concluding remarks

The nature of γδ T cells allows their expansion from peripheral blood of healthy donors and their employment as effector cells in immunotherapies targeting cancer, by transferring them to cancer patients. This T cell subset has demonstrated efficacy in effectively eliminating breast cancer stem cells (BCSC) with a phenotype akin to triple-negative breast cancer (TNBC). This success presents a potential avenue for treating challenging cancer types that are difficult to address due to the absence of cancer-specific antigens. The outlined method serves as a standardized protocol for assessing the cytotoxicity of various types of effector cells. It facilitates the examination of any tumor type, allowing for the utilization of cells from a cell line or from a patient as potential targets.

6. Notes

1. Beside this, there are many protocols that result in the expansion of different γδ T cell subpopulations, so it is worth to previously check which is the γδ T cell population of interest (Deniger et al., 2014; Ferry et al., 2022; Landin et al., 2021; Siegers et al., 2013).
2. γδ T cell expansion yields are very much donor dependent. A first idea of how the Vδ1 and Vδ2 populations will expand can be achieved by analyzing the populations by flow cytometry prior to inducing expansion
3. Suitable effector-to-target cell ratios and detection times should be established prior to assay
4. Including more measuring timepoints allows to establish better the killing dynamics, and might result crucial when comparing killing of different cell lines
5. Under the expanded cells we can find Vγ9δ2 γδ T cells, which recognize phosphoantigens. Thus, recognition and killing can be favored by adding 10 μM Zoledronate, which triggers the accumulation of phosphoantigens or hindered by introducing 25 μM Mevastatin, which blocks the mevalonate pathway. Utilizing these controls provides insight into the preferred killing mechanism of gamma delta T cells for each type of target cell
6. Supernatants of the cytotoxicity assay can be frozen and used for ELISA, in the determination of relevant cytokines in T cell effector function like IFNs and IL2

7. The assay can be incubated for a previously determined period (between 8 and 20 h, depending on the E:T ratio used) for an end-point measurement of specific cell killing via flow cytometry, e.g. including a cell death stain
8. Further T cell traits like viability, activation, degranulation, exhaustion can be determined via flow cytometry

Acknowledgments

I would like to acknowledge the excellent supervision and support of Prof. Susana Minguet Garcia, in whose laboratory this protocol has been written and the method performed. This work was supported by German Research Foundation (DFG) through SFB1479 (project ID: 441891347—P15 to S. Minguet).

Declaration of interests

The author has no financial conflict of interest.

References

Adams, E. J., Gu, S., & Luoma, A. M. (2015). Human gamma delta T cells: Evolution and ligand recognition. *Cellular Immunology, 296*(1), 31–40. https://doi.org/10.1016/j.cellimm.2015.04.008.

Andreu-Ballester, J. C., Galindo-Regal, L., Hidalgo-Coloma, J., Cuéllar, C., García-Ballesteros, C., Hurtado, C., et al. (2020). Differences in circulating γδ T cells in patients with primary colon cancer and relation with prognostic factors. *PLoS ONE, 15*(12), e0243545. https://doi.org/10.1371/journal.pone.0243545.

Boissière-Michot, F., Chabab, G., Mollevi, C., Guiu, S., Lopez-Crapez, E., Ramos, J., et al. (2021). Clinicopathological correlates of γδ T cell infiltration in triple-negative breast cancer. *Cancers, 13*(4), 765. https://doi.org/10.3390/cancers13040765.

Bryant, N. L., Suarez-Cuervo, C., Gillespie, G. Y., Markert, J. M., Nabors, L. B., Meleth, S., et al. (2009). Characterization and immunotherapeutic potential of γδ T-cells in patients with glioblastoma. *Neuro-Oncology, 11*(4), 357–367. https://doi.org/10.1215/15228517-2008-111.

Dang, A. T., Strietz, J., Zenobi, A., Khameneh, H. J., Brandl, S. M., Lozza, L., et al. (2021). NLRC5 promotes transcription of BTN3A1-3 genes and Vγ9Vδ2 T cell-mediated killing. *iScience, 24*(1), 101900. https://doi.org/10.1016/j.isci.2020.101900.

Deniger, D. C., Maiti, S. N., Mi, T., Switzer, K. C., Ramachandran, V., Hurton, L. V., et al. (2014). Activating and propagating polyclonal gamma delta T cells with broad specificity for malignancies. *Clinical Cancer Research, 20*(22), 5708–5719. https://doi.org/10.1158/1078-0432.CCR-13-3451.

Deseke, M., & Prinz, I. (2020). Ligand recognition by the γδ TCR and discrimination between homeostasis and stress conditions. *Cellular & Molecular Immunology, 17*(9), 914–924. https://doi.org/10.1038/s41423-020-0503-y.

Ferry, G. M., Agbuduwe, C., Forrester, M., Dunlop, S., Chester, K., Fisher, J., et al. (2022). A simple and robust single-step method for CAR-Vδ1 γδT cell expansion and transduction for cancer immunotherapy. *Frontiers in Immunology, 13*. https://doi.org/10.3389/fimmu.2022.863155.

Foord, E., Arruda, L. C. M., Gaballa, A., Klynning, C., & Uhlin, M. (2021). Characterization of ascites- and tumor-infiltrating γδ T cells reveals distinct repertoires and a beneficial role in ovarian cancer. *Science Translational Medicine, 13*(577). https://doi.org/10.1126/scitranslmed.abb0192.

Girardi, M., Oppenheim, D. E., Steele, C. R., Lewis, J. M., Glusac, E., Filler, R., et al. (2001). Regulation of cutaneous malignancy by γδ T cells. *Science, 294*(5542), 605–609. https://doi.org/10.1126/science.1063916.

Hartl, F. A., Beck-Garcìa, E., Woessner, N. M., Flachsmann, L. J., Cárdenas, R. M.-H. V., Brandl, S. M., et al. (2020). Noncanonical binding of Lck to CD3ε promotes TCR signaling and CAR function. *Nature Immunology, 21*(8), 902–913. https://doi.org/10.1038/s41590-020-0732-3.

Hu, Y., Hu, Q., Li, Y., Lu, L., Xiang, Z., Yin, Z., et al. (2023). γδ T cells: Origin and fate, subsets, diseases and immunotherapy. *Signal Transduction and Targeted Therapy, 8*(1), 434. https://doi.org/10.1038/s41392-023-01653-8.

Landin, A. M., Cox, C., Yu, B., Bejanyan, N., Davila, M., & Kelley, L. (2021). Expansion and enrichment of gamma-delta (γδ) T cells from apheresed human product. *Journal of Visualized Experiments, 175*. https://doi.org/10.3791/62622.

Mensurado, S., Blanco-Domínguez, R., & Silva-Santos, B. (2023). The emerging roles of γδ T cells in cancer immunotherapy. *Nature Reviews Clinical Oncology, 20*(3), 178–191. https://doi.org/10.1038/s41571-022-00722-1.

Metzger, E., Stepputtis, S. S., Strietz, J., Preca, B.-T., Urban, S., Willmann, D., et al. (2017). KDM4 inhibition targets breast cancer stem–like cells. *Cancer Research, 77*(21), 5900–5912. https://doi.org/10.1158/0008-5472.CAN-17-1754.

Nishimoto, K. P., Barca, T., Azameera, A., Makkouk, A., Romero, J. M., Bai, L., et al. (2022). Allogeneic CD20-targeted γδ T cells exhibit innate and adaptive antitumor activities in preclinical B-cell lymphoma models. *Clinical & Translational Immunology, 11*(2). https://doi.org/10.1002/cti2.1373.

Raute, K., Strietz, J., Parigiani, M. A., Andrieux, G., Thomas, O. S., Kistner, K. M., et al. (2023). Breast cancer stem cell–derived tumors escape from γδ T-cell immunosurveillance in vivo by modulating γδ T-cell ligands. *Cancer Immunology Research, 11*(6), 810–829. https://doi.org/10.1158/2326-6066.CIR-22-0296.

Saura-Esteller, J., de Jong, M., King, L. A., Ensing, E., Winograd, B., de Gruijl, T. D., et al. (2022). Gamma delta T-cell based cancer immunotherapy: Past-present-future. *Frontiers in Immunology, 13*. https://doi.org/10.3389/fimmu.2022.915837.

Siegers, G. M., Ribot, E. J., Keating, A., & Foster, P. J. (2013). Extensive expansion of primary human gamma delta T cells generates cytotoxic effector memory cells that can be labeled with Feraheme for cellular MRI. *Cancer Immunology, Immunotherapy, 62*(3), 571–583. https://doi.org/10.1007/s00262-012-1353-y.

Smolarz, B., Nowak, A. Z., & Romanowicz, H. (2022). Breast cancer—Epidemiology, classification, pathogenesis and treatment (review of literature). *Cancers, 14*(10), 2569. https://doi.org/10.3390/cancers14102569.

Wang, J., Lin, C., Li, H., Li, R., Wu, Y., Liu, H., et al. (2017). Tumor-infiltrating γδT cells predict prognosis and adjuvant chemotherapeutic benefit in patients with gastric cancer. *Oncoimmunology, 6*(11), e1353858. https://doi.org/10.1080/2162402X.2017.1353858.

Zhao, N., Dang, H., Ma, L., Martin, S. P., Forgues, M., Ylaya, K., et al. (2021). Intratumoral γδ T-cell infiltrates, chemokine (C-C motif) ligand 4/chemokine (C-C motif) ligand 5 protein expression and survival in patients with hepatocellular carcinoma. *Hepatology, 73*(3), 1045–1060. https://doi.org/10.1002/hep.31412.

9780443218682